T3-BOH-510

WITHDRAWN

MATHEMATICAL TOPICS
General Editor: C. PLUMPTON

VECTORS AND MATRICES

VECTORS
AND
MATRICES

BY

PAMELA LIEBECK

PERGAMON PRESS

OXFORD · NEW YORK · TORONTO

SYDNEY · BRAUNSCHWEIG

Pergamon Press Ltd., Headington Hill Hall, Oxford
Pergamon Press Inc., Maxwell House, Fairview Park, Elmsford, New York 10523
Pergamon of Canada Ltd., 207 Queen's Quay West, Toronto 1
Pergamon Press (Aust.) Pty. Ltd., 19a Boundary Street,
Rushcutters Bay, N.S.W. 2011, Australia
Vieweg & Sohn GmbH, Burgplatz 1, Braunschweig

First edition 1971
Library of Congress Catalog Card No. 70-130369

Printed in Hungary

08 015822 6 (flexicover)
08 015823 4 (hard cover)

CONTENTS

PREFACE

THIS book is intended for readers who require an unsophisticated approach to vectors and matrices which they may read on their own or use in a classroom situation. Such readers may include teachers of "modern" mathematics who wish to put their work with children into a wider framework, prospective university students wishing for a bridge between school and university algebra, and students at a college of education. I have presented the material over a number of years in lecture courses to such students and such teachers.

The approach to the subject is progressive, each concept being firmly established on an intuitive basis before being treated formally. Worked examples are an integral part of the text, providing illustrations and motivation for ensuing lines of thought. The basic knowledge required of the reader is a little two-dimensional coordinate geometry. The emphasis in the first half of the book is on geometry. Matrices are introduced through their association with geometry mappings. The second half of the book shows the importance of matrices in non-geometric situations, such as the theory of linear equations and eigenvector theory. Throughout the book emphasis is laid on the usefulness of the new concepts as well as on their mathematical structure. Chapter 10 in particular shows some of the power of eigenvector theory, applying it to some easily understood problems in biology, probability and genetics.

I should like to express gratitude to Dr. Plumpton of Queen Mary College, London, and to Mrs. P. Ducker of Pergamon Press for their helpful advice. Thanks are also due to my husband, Dr. H. Liebeck of Keele University, for many helpful discussions; to Mrs. M. Pratt for reading the manuscript and commenting from the point of view of a person with no advanced mathematical knowledge; and to my children, without whose occasional absence this book would never have been written.

PAMELA LIEBECK

CHAPTER 1

VECTORS AND SCALARS

Introduction

The simplest-sounding questions are often the hardest to answer. One such question is "What is a number?" It is easier to list the various types of numbers and the various properties of numbers than to define the actual numbers themselves. We shall pursue such a discussion before proceeding to the main work of this chapter, which will involve discussing in a similar way the quantities called vectors.

Types of Numbers

These are listed here for reference purposes.

Natural numbers are the numbers 1, 2, 3, 4, ..., etc.

Integers include the natural numbers, their negative counterparts and the number 0, or zero. One visual representation of the integers is the set of markings made on one of the axes for drawing graphs in coordinate geometry. Such an axis, illustrated in Fig. 1.1, is sometimes called a number line.

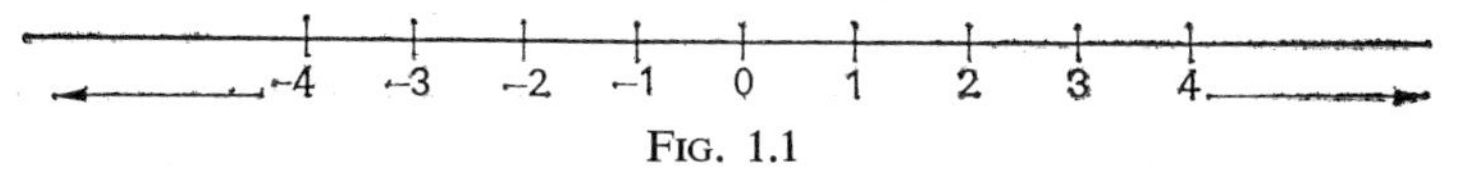

FIG. 1.1

Rational numbers may be expressed as a/b, where a and b are integers and b is not zero. These numbers could be represented visually by further markings on the number line of Fig. 1.1.

Irrational numbers are those which cannot be expressed as the ratio of two integers. $\sqrt{2}$ is such a number. $\sqrt{2}$ may be represented visually by the hypotenuse of a right-angled triangle with two of its sides one unit in length, as in Fig. 1.2.

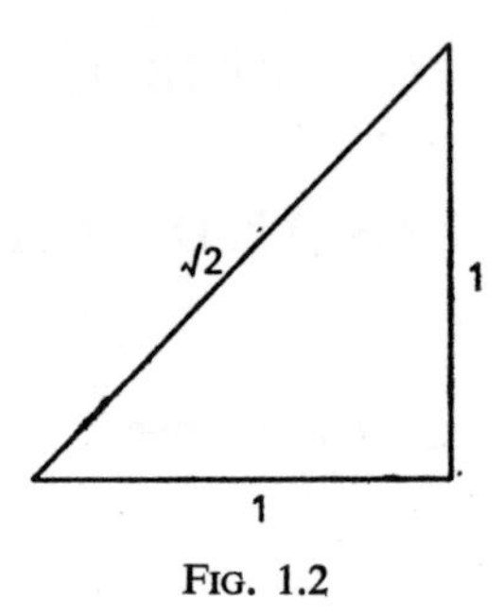

FIG. 1.2

Real numbers include all the numbers mentioned so far.

Imaginary numbers, so called because they seem so remote from visual aspects, are those which are the square root of a negative real number, such as $\sqrt{-1}$, usually denoted by i, or $\sqrt{-9}$, which is $3i$.

Complex numbers are numbers of the form $a+ib$, where a and b are real numbers. It will be seen that all the numbers mentioned so far are examples of complex numbers.

In the foregoing, the link between numbers and geometry has been stressed. This link is often illuminating. Sometimes algebra illuminates geometry, sometimes vice versa. We shall meet many instances in this book.

Properties of Numbers

We tend to take for granted not only what numbers are, but what are the operations we perform between them, such as addition and multiplication. It is easier to discuss the properties of these operations than to define the operations themselves. The following properties, you will observe, hold for all types of numbers mentioned.

$$\begin{aligned}
&\text{(i)} & a+b &= b+a,\\
&\text{(ii)} & a\times b &= b\times a,\\
&\text{(iii)} & a+(b+c) &= (a+b)+c,\\
&\text{(iv)} & a\times(b\times c) &= (a\times b)\times c,\\
&\text{(v)} & a\times(b+c) &= (a\times b)+(a\times c).
\end{aligned}$$

(i) and (ii) express the *commutative properties* of addition and multiplication, (iii) and (iv) the *associative properties* of addition and multiplication, and (v) the *distributive property* of multiplication over addition. We use these properties continually in our mental calculations with numbers. For example, to calculate $3+12$, we probably conceive the 12 first and mentally add 3, using the property (i). Or, consider the summing of $(7+18)+2$. The brackets indicate that 2 is to be added to the sum of 7 and 18. If instead we add 7 to the sum of 18 and 2, we are using the property (iii). Lastly, consider the calculation 21×5. If we add 20×5 to 1×5, we are using the property (v).

Exercise 1a

1. Think up everyday meanings of the words "rational", "irrational", and related words such as "ratio". Look for connections with the mathematical meanings.

2. Which is nearer to $\sqrt{2}$: 1·4 or 1·5? Find a rational number that is within 0·01 of $\sqrt{2}$. Look up a proof that $\sqrt{2}$ is not a rational number. (*What is Mathematics?*, by Courant and Robbins, or any book on the theory of numbers.)

3. Find (i) all integers, (ii) all real numbers, that may be substituted for x in each of the following equations:

(a) $x+3 = 0$	(c) $3x = 0$	(e) $4x^2 = 16$
(b) $0x = 0$	(d) $3x = 1$	(f) $16x^2 = 4$.

4. Which of the properties of numbers (i) to (v) hold for negative numbers?

5. Think up quick ways of doing the following calculations and decide which of the five properties of numbers you are using in each case:

(a) $(19\times 2)\times 5$	(c) 33×3
(b) $3+49$	(d) 29×3.

Did you use negative numbers in any of your calculations?

6. Select individual children whom you know, of varying ages or abilities, and test which of the properties (i) to (v) they have so far discovered.

Vectors

Geometrically, a vector quantity is completely specified by a numerical value and a direction. One example of such a quantity is a displacement, which is completely specified by a distance moved and the direction of motion. Another example is a velocity, which is completely specified by a numerical value of speed and the direction of motion. (Two cars with the same speedometer reading travelling in different directions have the same speed but different velocities.) Further examples of vector quantities include force and momentum, but these lie outside the scope of this book.

Notation. **Bold type** will be used to denote vector quantities. The vector $\mathbf{v}$ may be represented geometrically by a line segment OP, whose length represents the numerical value of $\mathbf{v}$ and whose direction represents that of $\mathbf{v}$ (see Fig. 1.3). We shall use the further notation $\overline{OP}$ to denote the vector $\mathbf{v}$, so that $\mathbf{v} = \overline{OP}$.

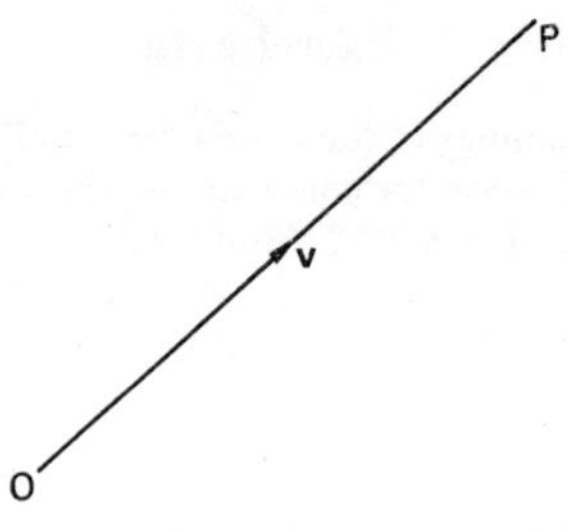

FIG. 1.3

Coordinates which are real numbers may also be used to specify a vector quantity. For example, referred to cartesian axes Ox, Oy in a plane which we shall call the x–y-plane, $\mathbf{v}_1 = \begin{pmatrix} 3 \\ 3 \end{pmatrix}$ is a coordinate vector representing a displacement of 3 units parallel to Ox and 3 units parallel to Oy. Figure 1.4 shows several representations of $\mathbf{v}_1$ and of $\mathbf{v}_2 = \begin{pmatrix} -2 \\ 0 \end{pmatrix}$.

There are infinitely many possible representations of $\mathbf{v}_1$ and $\mathbf{v}_2$, as the coordinates represent a vector and not a point as in coordinate geometry. A vector is specified by size and direction, but not necessarily by position.

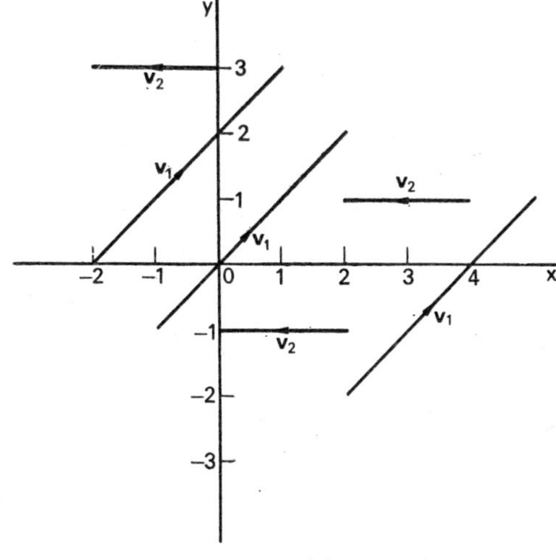

FIG. 1.4

Scalars. To distinguish them from vector quantities, those quantities that are specified by a numerical value only are often called scalar quantities. For example, distance and speed are scalar quantities, whereas displacement and velocity are vector quantities.

Addition of Vectors

A displacement from O to P followed by a displacement from P to Q has the same end result as a displacement straight from O to Q (see Fig. 1.5). The "following on" of displacements is an example of vector addition.

DEFINITION 1.1. *The sum of two vectors* $\overline{OP}$ *and* $\overline{PQ}$ *is the vector* $\overline{OQ}$, *where* OQ *completes the triangle* OPQ. *For example, if* $\overline{OP} = \mathbf{v}$, $\overline{PQ} = \mathbf{w}$, *then* $\overline{OQ} = \mathbf{v}+\mathbf{w}$.

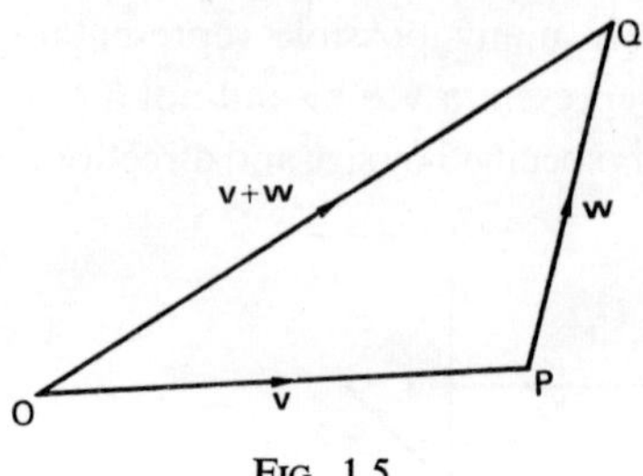

FIG. 1.5

Note. This is a different use of the word "sum" and the symbol "+" from those which pertain to scalar quantities. $OP+PQ = OQ$ is a statement about lengths, and it is a false statement unless OPQ is a straight line. But $\overline{OP}+\overline{PQ} = \overline{OQ}$ is a statement about vectors, using the old symbol "+" with a new meaning.

In the examples that follow, the sketches in the text will apply to part (a) of each example. Try to supply the sketches that apply to part (b) in each case.

EXAMPLE (i). $\mathbf{v}_1$ and $\mathbf{v}_2$ have been specified and illustrated in Fig. 1.4. Specify and illustrate the vectors (a) $\mathbf{v}_1+\mathbf{v}_2$, (b) $\mathbf{v}_2+\mathbf{v}_1$.

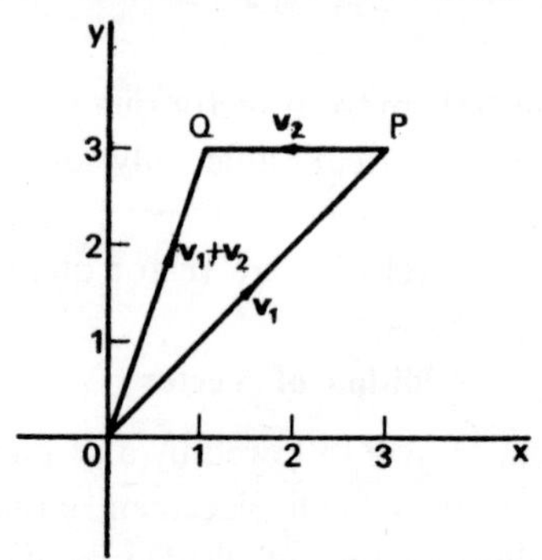

FIG. 1.6

In Fig. 1.6 we select a representation $\overline{OP}$ of $\mathbf{v}_1$ and a representation $\overline{PQ}$ of $\mathbf{v}_2$ which follows on from $\mathbf{v}_1$. $\overline{OQ}$ will then be a representation of $\mathbf{v}_1+\mathbf{v}_2$, and its specification is $\begin{pmatrix}1\\3\end{pmatrix}$.

The illustration for (b) will differ from that for (a), but it should show that $\mathbf{v}_2+\mathbf{v}_1 = \begin{pmatrix}1\\3\end{pmatrix}$.

Example (i) illustrates the *commutative property* of vector addition, that for any two vectors $\mathbf{v}$ and $\mathbf{w}$,

$$\mathbf{v}+\mathbf{w} = \mathbf{w}+\mathbf{v}.$$

EXAMPLE (ii). Specify and illustrate the vectors (a) $\mathbf{v}_1+(\mathbf{v}_2+\mathbf{v}_3)$, (b) $(\mathbf{v}_1+\mathbf{v}_2)+\mathbf{v}_3$, where $\mathbf{v}_3 = \begin{pmatrix}-3\\1\end{pmatrix}$.

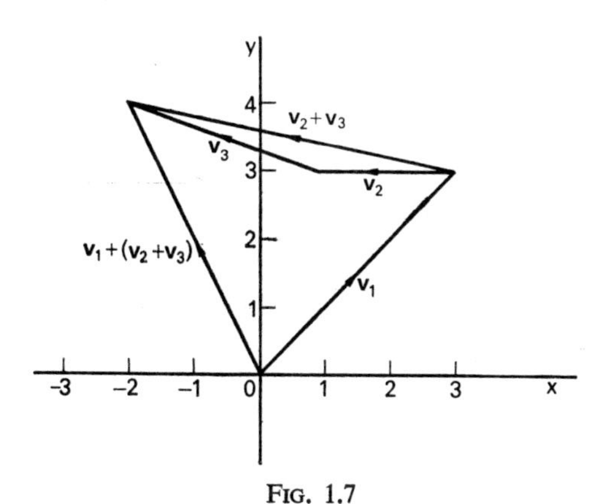

FIG. 1.7

In Fig. 1.7 we select a representation of $(\mathbf{v}_2+\mathbf{v}_3)$ that follows on from $\mathbf{v}_1$, and its specification is $\begin{pmatrix}-2\\4\end{pmatrix}$.

The illustration for (b) will differ from that for (a), but it should show that $(\mathbf{v}_1+\mathbf{v}_2)+\mathbf{v}_3 = \begin{pmatrix}-2\\4\end{pmatrix}$.

Example (ii) illustrates the *associative property* of vector addition, that for any three vectors $\mathbf{u}$, $\mathbf{v}$ and $\mathbf{w}$,

$$\mathbf{u}+(\mathbf{v}+\mathbf{w}) = (\mathbf{u}+\mathbf{v})+\mathbf{w}.$$

Addition Rule for Coordinate Vectors

The rule $$\begin{pmatrix}x_1\\y_1\end{pmatrix}+\begin{pmatrix}x_2\\y_2\end{pmatrix}=\begin{pmatrix}x_1+x_2\\y_1+y_2\end{pmatrix}$$
can be seen to hold for Examples (i) and (ii). For instance, in Example (i) we had: $\begin{pmatrix}3\\3\end{pmatrix}+\begin{pmatrix}-2\\0\end{pmatrix}=\begin{pmatrix}1\\3\end{pmatrix}$. That the rule is generally true can be shown by applying the geometric properties of the rectangle to Fig. 1.8.

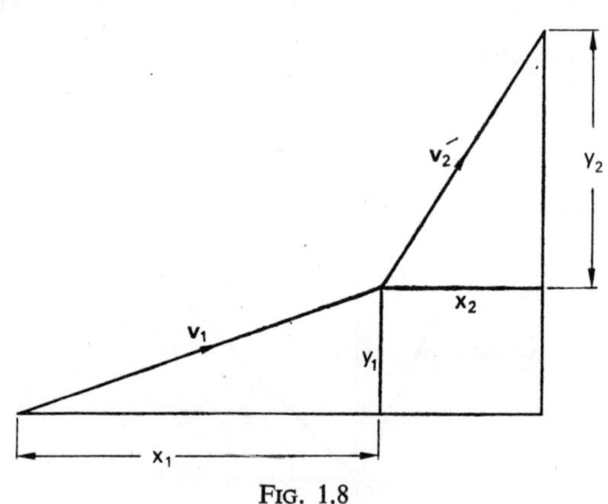

FIG. 1.8

Zero and Negative Vectors

Like the number 0, the zero vector quantity, denoted by **0**, is an essential concept. With vector addition already defined, the zero vector will have to satisfy

$$\mathbf{v}+\mathbf{0}=\mathbf{v} \quad \text{for any vector } \mathbf{v}.$$

DEFINITION 1.2. *The zero vector* **0** *is a vector whose size is zero. It is meaningless to associate a direction with the vector* **0**.

It follows that $\mathbf{v}+\mathbf{0}=\mathbf{v}$ for any vector **v**, and that the coordinate vector $\mathbf{0}=\begin{pmatrix}0\\0\end{pmatrix}$.

DEFINITION 1.3. *The negative vector* $-\mathbf{v}$ *has the same size as* $\mathbf{v}$, *but a direction opposite to that of* $\mathbf{v}$.

It follows that $\mathbf{v}+(-\mathbf{v}) = \mathbf{0}$ for any vector $\mathbf{v}$, and that if $\mathbf{v}$ is the coordinate vector $\begin{pmatrix} x \\ y \end{pmatrix}$, then the vector $-\mathbf{v} = \begin{pmatrix} -x \\ -y \end{pmatrix}$.

EXAMPLE (iii). Specify $\mathbf{v}_1+\mathbf{v}_4$, where $\mathbf{v}_4 = \begin{pmatrix} -3 \\ -3 \end{pmatrix}$.

By the addition rule, $\mathbf{v}_1+\mathbf{v}_4 = \begin{pmatrix} 3 \\ 3 \end{pmatrix} + \begin{pmatrix} -3 \\ -3 \end{pmatrix} = \begin{pmatrix} 0 \\ 0 \end{pmatrix} = \mathbf{0}$.

Notice that $\mathbf{v}_1 = -\mathbf{v}_4$, and $\mathbf{v}_4 = -\mathbf{v}_1$.

EXAMPLE (iv). Specify and illustrate (a) $\mathbf{v}_1+(-\mathbf{v}_2)$, (b) $\mathbf{v}_2+(-\mathbf{v}_1)$.

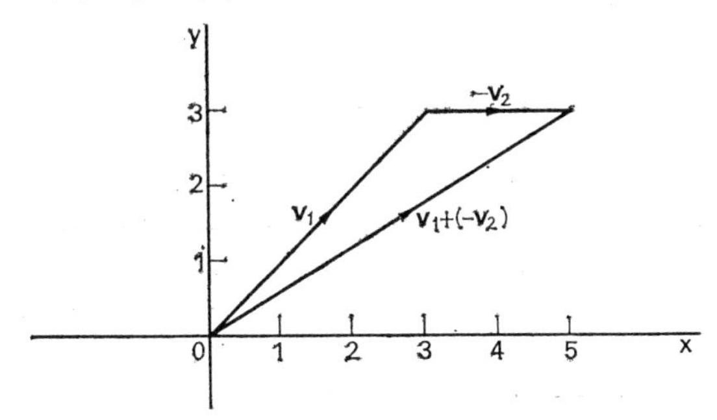

FIG. 1.9

In Fig. 1.9 we construct a representation of $-\mathbf{v}_2$, in a direction opposite to $\mathbf{v}_2$, to follow on from $\mathbf{v}_1$.

By the addition rule, (a) $\mathbf{v}_1+(-\mathbf{v}_2) = \begin{pmatrix} 3 \\ 3 \end{pmatrix} + \begin{pmatrix} 2 \\ 0 \end{pmatrix} = \begin{pmatrix} 5 \\ 3 \end{pmatrix}$,

(b) $\mathbf{v}_2+(-\mathbf{v}_1) = \begin{pmatrix} -2 \\ 0 \end{pmatrix} + \begin{pmatrix} -3 \\ -3 \end{pmatrix} = \begin{pmatrix} -5 \\ -3 \end{pmatrix}$.

Notation. With numbers it is convenient to write "5−3" to denote "5+(−3)". Similarly, with vectors, we may write "$\mathbf{v}-\mathbf{w}$" to denote "$\mathbf{v}+(-\mathbf{w})$".

Check in Example (iv) that $\mathbf{v}_1-\mathbf{v}_2 = -(\mathbf{v}_2-\mathbf{v}_1)$.

Relevance of Vector Addition

We noted that the vector sum of two displacements represents the result of one displacement followed by the other. Relevance can be attached to the sum of other vector quantities. For instance, the vector sum of two velocities represents the resulting velocity of an object that has those two velocities simultaneously. (An aeroplane, for example, may propel itself in one direction while the wind blows it in another.) Indeed, the concept of vector sum has importance for any type of vector quantity.

The zero vector has obvious significance in terms of displacement or velocity. So does a negative vector; for example, if $\mathbf{v}$ denotes a velocity of 60 miles per hour in a northerly direction, then $-\mathbf{v}$ denotes a velocity of 60 miles per hour in a southerly direction.

Product of Scalar and Vector

We now consider the possibility of operations between scalars and vectors. Since addition has different meanings when applied to the two types of quantities, addition of scalars and vectors is meaningless. But there is a sensible interpretation of the product of scalar and vector.

DEFINITION 1.4. *We define* $k\mathbf{v}$, *for* k:

(a) *a positive real number,*
(b) *a negative real number,*
(c) *zero.*

For case (a), $k\mathbf{v}$ *is a vector having the same direction as* $\mathbf{v}$ *and* k *times the size of* $\mathbf{v}$.
For case (b), $k\mathbf{v}$ *is a vector having a direction opposite to* $\mathbf{v}$ *and* $|k|$† *times the size of* $\mathbf{v}$.
For case (c), $k\mathbf{v}$ *is the zero vector.*
For all cases, $\mathbf{v}k$ *has the same meaning as* $k\mathbf{v}$.

† For real numbers k, $|k| = +k$ or $-k$, according as k is non-negative or negative. The quantity $|k|$ is called the *absolute value* of k.

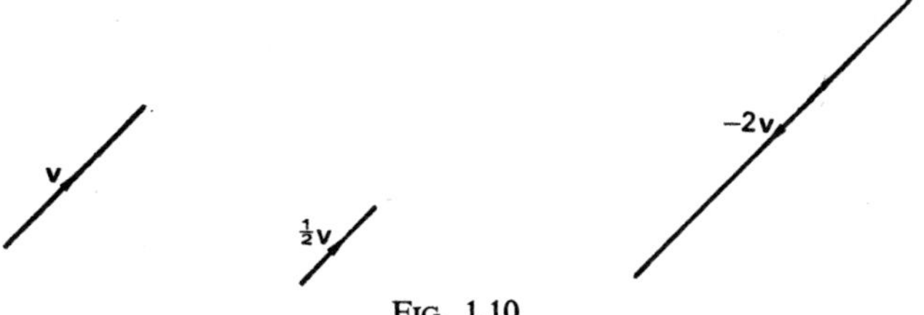

FIG. 1.10

EXAMPLE (v). Specify and illustrate (a) $2\mathbf{v}_1$, (b) $2\mathbf{v}_2$.

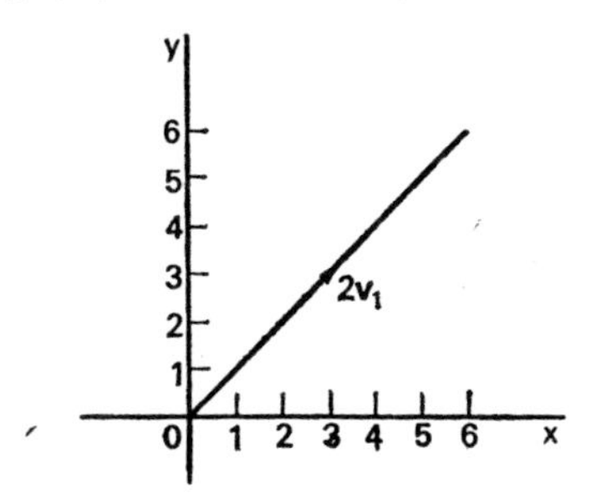

FIG. 1.11

From Fig. 1.11, we see that $2\mathbf{v}_1 = \begin{pmatrix}6\\6\end{pmatrix}$.

The illustration for (b) should show that $2\mathbf{v}_2 = \begin{pmatrix}-4\\0\end{pmatrix}$.

Rule for Product of Scalar and Coordinate Vector

The rule
$$k\begin{pmatrix}x\\y\end{pmatrix} = \begin{pmatrix}kx\\ky\end{pmatrix}$$
can be seen to hold for Example (v), and it can be shown to be generally true by applying the properties of similar triangles to Fig. 1.12.

EXAMPLE (vi). Specify (a) $2\mathbf{v}_1+2\mathbf{v}_2$, (b) $2(\mathbf{v}_1+\mathbf{v}_2)$.

(a) $2\mathbf{v}_1+2\mathbf{v}_2 = 2\begin{pmatrix}3\\3\end{pmatrix}+2\begin{pmatrix}-2\\0\end{pmatrix} = \begin{pmatrix}6\\6\end{pmatrix}+\begin{pmatrix}-4\\0\end{pmatrix} = \begin{pmatrix}2\\6\end{pmatrix}$.

(b) $2(\mathbf{v}_1+\mathbf{v}_2) = 2\left[\begin{pmatrix}3\\3\end{pmatrix}+\begin{pmatrix}-2\\0\end{pmatrix}\right] = 2\begin{pmatrix}1\\3\end{pmatrix} = \begin{pmatrix}2\\6\end{pmatrix}$.

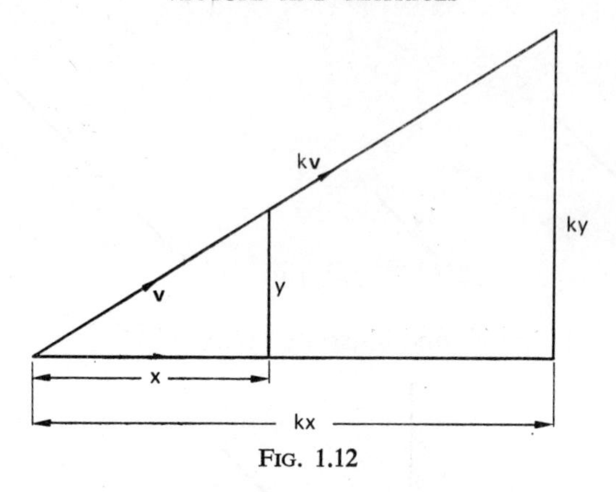

FIG. 1.12

Example (vi) illustrates the *distributive property*, that for any two vectors **v** and **w**, and any scalar quantity k,

$$k(\mathbf{v}+\mathbf{w}) = k\mathbf{v}+k\mathbf{w}.$$

Try to prove this property for coordinate vectors by combining the two rules we have established for coordinate vectors.

In the final example of this chapter, we shall use the properties we have established to prove a theorem in plane geometry.

EXAMPLE (vii). Prove that the diagonals of a parallelogram bisect each other.

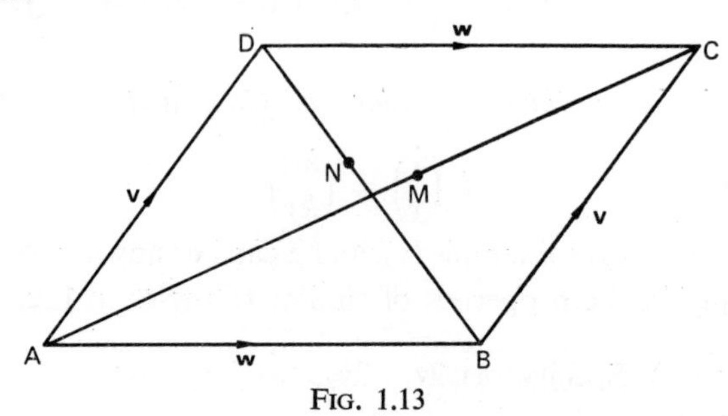

FIG. 1.13

Referring to Fig. 1.13, we shall prove that the mid-point, M, of AC coincides with the mid-point, N, of BD. **v** denotes the vector $\overline{AD}$, **w** the vector $\overline{AB}$.

$$
\begin{aligned}
\overline{AC} &= \mathbf{v}+\mathbf{w}, && \text{by Definition 1.1.}\\
\overline{AM} &= \tfrac{1}{2}\overline{AC}, && \text{by Definition 1.4(a),}\\
&= \tfrac{1}{2}(\mathbf{v}+\mathbf{w}), && \\
&= \tfrac{1}{2}\mathbf{v}+\tfrac{1}{2}\mathbf{w}, && \text{by the distributive property.}\\
\overline{BD} &= \mathbf{v}-\mathbf{w}, && \text{by Definitions 1.1, 1.3.}\\
\overline{BN} &= \tfrac{1}{2}(\mathbf{v}-\mathbf{w}), && \text{by Definition 1.4(a),}\\
&= \tfrac{1}{2}\mathbf{v}-\tfrac{1}{2}\mathbf{w}, && \text{by the distributive property.}\\
\overline{AN} &= \overline{AB}+\overline{BN}, && \text{by Definition 1.1,}\\
&= \mathbf{w}+(\tfrac{1}{2}\mathbf{v}-\tfrac{1}{2}\mathbf{w}), && \\
&= \tfrac{1}{2}\mathbf{v}+\tfrac{1}{2}\mathbf{w}, && \text{by distributive and associative properties.}
\end{aligned}
$$

Thus AN and AM have the same size and direction, and M and N coincide.

Summary of Chapter 1

We have defined two operations: vector addition and multiplication of scalar and vector.

Two rules for coordinate vectors emerge from the definitions:

(1) Addition Rule: $\begin{pmatrix} x_1 \\ y_1 \end{pmatrix} + \begin{pmatrix} x_2 \\ y_2 \end{pmatrix} = \begin{pmatrix} x_1+x_2 \\ y_1+y_2 \end{pmatrix}$,

(2) Scalar times Vector Rule: $k\begin{pmatrix} x \\ y \end{pmatrix} = \begin{pmatrix} kx \\ ky \end{pmatrix}$.

Three properties of these operations emerge:

(1) the commutative property of vector addition,
(2) the associative property of vector addition,
(3) a distributive property.

Exercise 1b

1. Which of the following statements are meaningless?

(a) $\mathbf{v} = 7+\mathbf{w}$ (c) $\mathbf{v}+\mathbf{w} = 0$ (e) $\mathbf{v}-\mathbf{w} = 7$
(b) $\mathbf{v} = 7\mathbf{w}$ (d) $\mathbf{v}+\mathbf{w} = \mathbf{0}$ (f) $\mathbf{v}-\mathbf{w} = 7\mathbf{w}$.

2. $\mathbf{p} = \begin{pmatrix}1\\0\end{pmatrix}$, $\mathbf{q} = \begin{pmatrix}0\\-3\end{pmatrix}$, $\mathbf{r} = \begin{pmatrix}-2\\2\end{pmatrix}$. Specify and illustrate the vectors **v**, **w**, **u**, where:

$$\mathbf{v} = \mathbf{p}+\mathbf{q}, \quad \mathbf{w} = \mathbf{q}-\mathbf{r}, \quad \mathbf{u} = \mathbf{v}-\mathbf{r}-\mathbf{p}.$$

You should find two of your results equal. Can you prove without reference to the co-ordinates that this must be so?

3. The points A and B are such that $\overline{OA} = \begin{pmatrix}3\\1\end{pmatrix}$, $\overline{OB} = \begin{pmatrix}2\\-1\end{pmatrix}$. $AOBP$ is a parallelogram. So are $ABOQ$ and $OABR$. Specify the vectors $\overline{PQ}$, $\overline{QR}$, $\overline{RP}$. What do you notice?

4. $ABCD$ is a quadrilateral. $\overline{AD} = \overline{BC} = \mathbf{v}$, $\overline{AB} = \mathbf{w}$. What type of quadrilateral is $ABCD$? Find $\overline{CD}$, $\overline{BD}$, $\overline{CA}$ in terms of **v** and **w**.

5. $ABCD$ is a quadrilateral. $\overline{AD} = 3\mathbf{v}$, $\overline{BC} = 2\mathbf{v}$, $\overline{AB} = \mathbf{w}$. What type of quadrilateral is $ABCD$? Find $\overline{CD}$, $\overline{BD}$, $\overline{CA}$ in terms of **v** and **w**.

6. The wind blows from the east at 50 m.p.h. A plane flies on a compass bearing due north at a speedometer reading of 120 m.p.h. What is the plane's velocity as observed from the ground?

7. In triangle ABC, L is the mid-point of AB, M the mid-point of AC. Use a vector method to prove that ML, BC are parallel.

8. In triangle ABC, M is the mid-point of BC. $\overline{AB} = \mathbf{c}$, $\overline{AC} = \mathbf{b}$. Find $\overline{AM}$ in terms of **b** and **c**. (Example (vii) should help.)

   P and Q trisect BC. Find $\overline{AP}$ and $\overline{AQ}$ in terms of **b** and **c**.

9. Prepare an introduction to vectors for 10-year-olds or for 15-year-olds. Think up suitable games or applications for the age group you have in mind. (See the Nuffield Project Guide on Vectors.)

CHAPTER 2

THE INNER PRODUCT

Introduction

In Chapter 1 we gave two basic definitions whose consequences served as tools for solving some modest problems in geometry. In this chapter we shall develop some more powerful tools for problem solving, based on what will initially seem a rather strange and arbitrary definition of a product of two vectors.

Notation. The size of the vector $\mathbf{v}$ is written v. So in Fig. 2.1, $v = w$ but $\mathbf{v} \neq \mathbf{w}$.

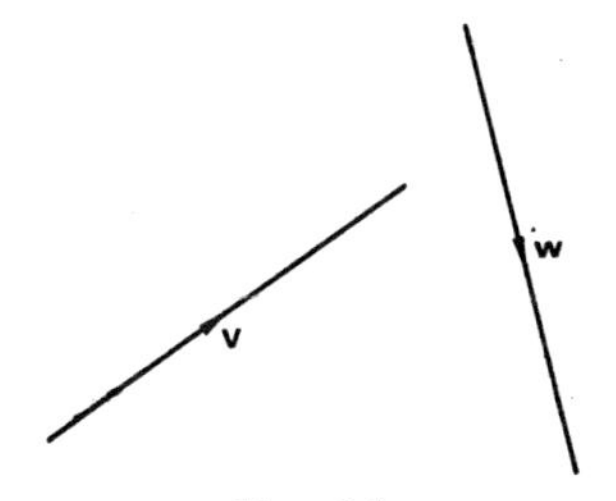

FIG. 2.1

EXAMPLE (i). $\mathbf{v}_1 = \begin{pmatrix} 3 \\ 3 \end{pmatrix}$ and $\mathbf{v}_2 = \begin{pmatrix} -2 \\ 0 \end{pmatrix}$. Find (a) v_1, (b) v_2.

(a) With reference to Fig. 2.2,

$$v_1^2 = 9+9, \quad \text{by Pythagoras' theorem.}$$

Thus $v_1 = 3\sqrt{2}$.

(b) $v_2 = 2$. (Note that v_2 is not -2. No vector has negative size.)

DEFINITION 2.1. *The angle between two vectors is that angle, not reflex, which is formed between representations of the vectors radiating from the same point.*

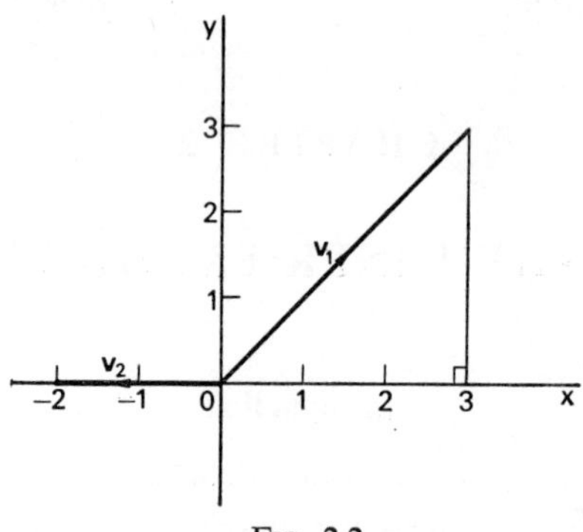

FIG. 2.2

EXAMPLE (ii). Find the angle between (a) $\mathbf{v}_1$ and $\mathbf{v}_2$, (b) $\mathbf{v}_1$ and $-\mathbf{v}_2$.

(a) In Fig. 2.2 we have representations of the vectors radiating from the same point, illustrating that the angle between $\mathbf{v}_1$ and $\mathbf{v}_2$ is 135°.

(b) $-\mathbf{v}_2$ would point in the opposite direction to $\mathbf{v}_2$, so the angle between $\mathbf{v}_1$ and $-\mathbf{v}_2$ is 45°.

Inner Product

DEFINITION 2.2. *The inner product of two vectors* $\mathbf{v}$ *and* $\mathbf{w}$ *is the scalar quantity* $vw \cos \alpha$, *where* α *is the angle between* $\mathbf{v}$ *and* $\mathbf{w}$.

Notice that in this definition we have the product of two vectors being a scalar quantity. The word "product" is used here (as "sum" was in Chapter 1) with a meaning quite different from that attached to numbers.

Notation. The symbol used for the inner product is a dot. In fact, the inner product is also known as the "dot product". We write:

$$\mathbf{v} \cdot \mathbf{w} = vw \cos \alpha.$$

EXAMPLE (iii). Calculate (a) $\mathbf{v}_1 \cdot \mathbf{v}_2$, (b) $\mathbf{v}_2 \cdot \mathbf{v}_1$.

$$\text{(a) } \mathbf{v}_1 \cdot \mathbf{v}_2 = 3\sqrt{2} \cdot 2 \cos 135^\circ$$
$$= -6.$$

$$\text{(b) } \mathbf{v}_2 \cdot \mathbf{v}_1 = 2 \cdot 3\sqrt{2} \cos 135^\circ$$
$$= -6.$$

Example (iii) illustrates the *commutative property* of the inner product. Why is it meaningless to discuss an associative property?

EXAMPLE (iv). Calculate (a) $\mathbf{v}_1 \cdot \mathbf{v}_1$, (b) $\mathbf{v}_1 \cdot \mathbf{v}_5$, where $\mathbf{v}_5 = \begin{pmatrix} 3 \\ -3 \end{pmatrix}$.

$$\text{(a) } \mathbf{v}_1 \cdot \mathbf{v}_1 = 3\sqrt{2} \cdot 3\sqrt{2} \cos 0^\circ$$
$$= 18.$$

$$\text{(b) } \mathbf{v}_1 \cdot \mathbf{v}_5 = 3\sqrt{2} \cdot 3\sqrt{2} \cos 90^\circ$$
$$= 0.$$

Example (iv) illustrates two important properties:

The inner product of a vector $\mathbf{v}$ *with itself is* v^2.

The inner product of two perpendicular vectors is zero.

EXAMPLE (v). Calculate (a) $\mathbf{v}_1 \cdot (-\mathbf{v}_2)$, (b) $(-\mathbf{v}_1) \cdot (-\mathbf{v}_2)$.

$$\text{(a) } \quad \mathbf{v}_1 \cdot (-\mathbf{v}_2) = 3\sqrt{2} \cdot 2 \cos 45^\circ$$
$$= 6.$$

$$\text{(b) } (-\mathbf{v}_1) \cdot (-\mathbf{v}_2) = 3\sqrt{2} \cdot 2 \cos 135^\circ$$
$$= -6.$$

Example (v) illustrates that for the inner product minus signs behave as they do for the ordinary product of numbers. That is, for any vectors $\mathbf{v}$ and $\mathbf{w}$,

$$\mathbf{v} \cdot (-\mathbf{w}) = -(\mathbf{v} \cdot \mathbf{w}),$$
$$(-\mathbf{v}) \cdot (-\mathbf{w}) = \mathbf{v} \cdot \mathbf{w}.$$

EXAMPLE (vi). Calculate (a) $\mathbf{v}_2 \cdot (\mathbf{v}_1 + \mathbf{v}_5)$, (b) $\mathbf{v}_2 \cdot \mathbf{v}_1 + \mathbf{v}_2 \cdot \mathbf{v}_5$.

$$\text{(a)} \qquad \mathbf{v}_1 + \mathbf{v}_5 = \begin{pmatrix} 6 \\ 0 \end{pmatrix},$$

$$\text{so } \mathbf{v}_2 \cdot (\mathbf{v}_1 + \mathbf{v}_5) = 2 \cdot 6 \cos 180^\circ$$
$$= -12.$$

$$\text{(b) } \mathbf{v}_2 \cdot \mathbf{v}_1 + \mathbf{v}_2 \cdot \mathbf{v}_5 = -6 + 2 \cdot 3\sqrt{2} \cos 135^\circ$$
$$= -6 - 6$$
$$= -12.$$

Example (vi) illustrates a *distributive property*, that for any vectors $\mathbf{v}$ and $\mathbf{w}$,

$$\mathbf{w} \cdot (\mathbf{u} + \mathbf{v}) = \mathbf{w} \cdot \mathbf{u} + \mathbf{w} \cdot \mathbf{v}.$$

The proof of this property may be found at the end of this chapter. Its implications are that *we may manipulate brackets with techniques similar to those we use in the algebra of numbers.*

EXAMPLE (vii). Prove Pythagoras' theorem.

We refer to Fig. 2.3, in which triangle ABC is right-angled at C, and $\overline{BC} = \mathbf{a}$, $\overline{CA} = \mathbf{b}$, and $\overline{BA} = \mathbf{c}$.

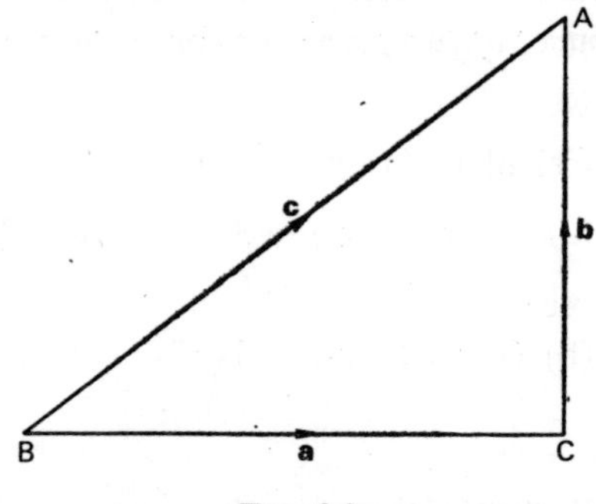

FIG. 2.3

Clearly $\qquad \mathbf{c} = \mathbf{a} + \mathbf{b},$

so $\qquad \mathbf{c} \cdot \mathbf{c} = (\mathbf{a} + \mathbf{b}) \cdot (\mathbf{a} + \mathbf{b})$

$$= \mathbf{a} \cdot \mathbf{a} + \mathbf{b} \cdot \mathbf{a} + \mathbf{a} \cdot \mathbf{b} + \mathbf{b} \cdot \mathbf{b}$$
$$= a^2 + 2\mathbf{a} \cdot \mathbf{b} + b^2.$$

But $\mathbf{a} \cdot \mathbf{b} = 0$, as **a** and **b** are perpendicular.

Thus
$$\mathbf{c} \cdot \mathbf{c} = a^2 + b^2,$$
or
$$c^2 = a^2 + b^2.$$

Note. If C is not a right angle, this argument leads to the cosine rule for triangles:

$$c^2 = a^2 + b^2 - 2ab \cos C.$$

Can you see why $2\mathbf{a} \cdot \mathbf{b} = -2ab \cos C$?

EXAMPLE (viii). If $v = w$, show that the vectors $\mathbf{v} - \mathbf{w}$ and $\mathbf{v} + \mathbf{w}$ are perpendicular.

We shall show that the inner product of the two vectors is zero:

$$\begin{aligned}(\mathbf{v} - \mathbf{w}) \cdot (\mathbf{v} + \mathbf{w}) &= \mathbf{v} \cdot \mathbf{v} - \mathbf{w} \cdot \mathbf{v} + \mathbf{v} \cdot \mathbf{w} - \mathbf{v} \cdot \mathbf{v} \\ &= v^2 - w^2 \\ &= 0.\end{aligned}$$

Figures 2.4 and 2.5 illustrate that this single vector property proves two geometry theorems. What are they?

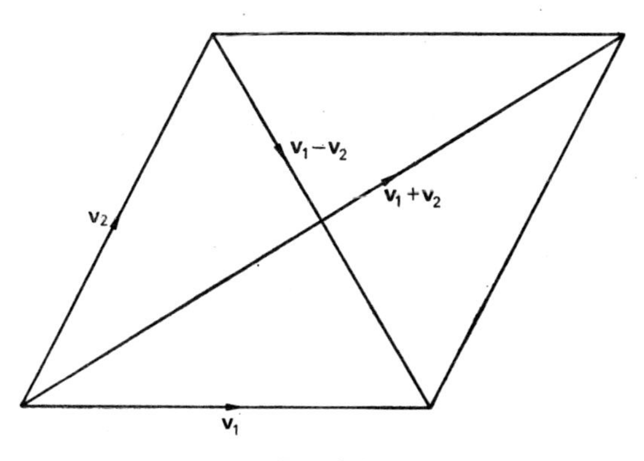

FIG. 2.4

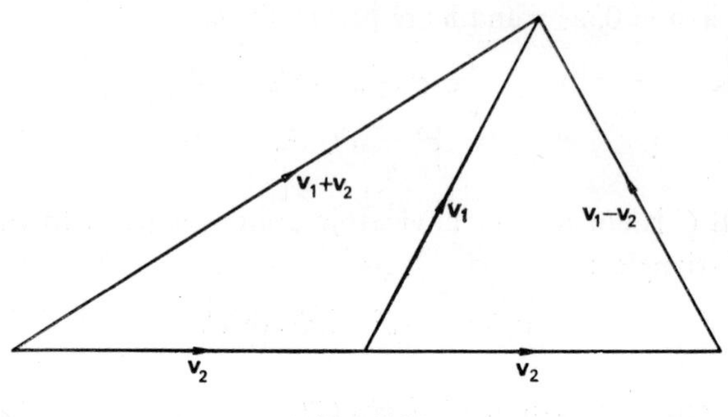

FIG. 2.5

Unit Vectors

We shall sometimes find it useful to denote by $\mathbf{i}$, $\mathbf{j}$, the vectors in the directions Ox, Oy, which have unit size.

That is,

$$\mathbf{i} = \begin{pmatrix}1\\0\end{pmatrix}, \quad \mathbf{j} = \begin{pmatrix}0\\1\end{pmatrix}.$$

Thus
$$\mathbf{i}\cdot\mathbf{i} = \mathbf{j}\cdot\mathbf{j} = 1, \quad \mathbf{i}\cdot\mathbf{j} = \mathbf{j}\cdot\mathbf{i} = 0,$$

and for any vector $\mathbf{v}$ with coordinates x and y,

$$\mathbf{v} = x\mathbf{i}+y\mathbf{j}.$$

Rule for Inner Product of Coordinate Vectors

Let $\mathbf{v} = x_1\mathbf{i}+y_1\mathbf{j}$, $\mathbf{w} = x_2\mathbf{i}+y_2\mathbf{j}$. We express $\mathbf{v}\cdot\mathbf{w}$ in terms of x_1, y_1, x_2, y_2.

$$\begin{aligned}\mathbf{v}\cdot\mathbf{w} &= (x_1\mathbf{i}+y_1\mathbf{j})\cdot(x_2\mathbf{i}+y_2\mathbf{j})\\ &= x_1x_2\mathbf{i}\cdot\mathbf{i}+y_1x_2\mathbf{j}\cdot\mathbf{i}+x_1y_2\mathbf{i}\cdot\mathbf{j}+y_1y_2\mathbf{j}\cdot\mathbf{j}\\ &= x_1x_2+y_1y_2.\end{aligned}$$

In obtaining this result, we have used the commutative and distributive properties of the inner product.

Note. If $\mathbf{v} = \mathbf{w}$ in the foregoing, we have

$$\mathbf{v} \cdot \mathbf{v} = v^2 = x_1^2 + y_1^2,$$

and have again proved Pythagoras' theorem.

Now check examples (i) to (vi) using this new rule for the inner product of coordinate vectors.

EXAMPLE (ix). Find vectors perpendicular to (a) $\mathbf{v}_1$, (b) $\mathbf{v}_2$.

(a) We require a vector $\mathbf{v} = \begin{pmatrix} x \\ y \end{pmatrix}$ such that $\mathbf{v}_1 \cdot \mathbf{v} = 0$.

Using the new rule, we have $3x + 3y = 0$.

Suitable values for $\mathbf{v}$ are thus $\begin{pmatrix} 1 \\ -1 \end{pmatrix}$, $\begin{pmatrix} -3 \\ 3 \end{pmatrix}$, etc. Notice that all suitable vectors are parallel.

(b) Clearly, any vector parallel to $\mathbf{j}$ is perpendicular to $\mathbf{v}_2$.

EXAMPLE (x). Find the angles α, β between (a) $\mathbf{v}_3 = \begin{pmatrix} -3 \\ 1 \end{pmatrix}$ and Ox, (b) $\mathbf{v}_3$ and $\mathbf{v}_1$.

(a) Using the rule, $\mathbf{v}_3 \cdot \mathbf{i} = -3 \cdot 1 + 1 \cdot 0$
$= -3.$

But also $\mathbf{v}_3 \cdot \mathbf{i} = \sqrt{10} \cdot 1 \cos \alpha.$

Thus $\cos \alpha = \dfrac{-3}{\sqrt{10}}.$

(b) Using the rule, $\mathbf{v}_3 \cdot \mathbf{v}_1 = -3 \cdot 3 + 1 \cdot 3$
$= -6.$

But also $\mathbf{v}_3 \cdot \mathbf{v}_1 = \sqrt{10} \cdot 3\sqrt{2} \cos \beta.$

Thus $\cos \beta = \dfrac{-2}{\sqrt{20}}.$

Example (x) (a) illustrates that for any vector $\mathbf{v}$, $\mathbf{v} \cdot \mathbf{i}$ is equal to the x coordinate of $\mathbf{v}$ (or the *projection* of $\mathbf{v}$ on Ox).

EXAMPLE (xi). $\overline{OV_1} = \mathbf{v} = \begin{pmatrix} x_1 \\ y_1 \end{pmatrix}$, $\overline{OV_2} = \mathbf{w} = \begin{pmatrix} x_2 \\ y_2 \end{pmatrix}$. Show that the area of the triangle OV_1V_2 is $\frac{1}{2}|x_1y_2 - x_2y_1|$.

Let angle $V_1OV_2 = \theta$. Then $\mathbf{v}\cdot\mathbf{w} = vw\cos\theta = x_1x_2+y_1y_2$.

Thus $$\cos\theta = \frac{x_1x_2+y_1y_2}{vw}.$$

Now $$\sin^2\theta = 1-\cos^2\theta$$
$$= \frac{(x_1^2+y_1^2)(x_2^2+y_2^2)-(x_1x_2+y_1y_2)^2}{v^2w^2}$$
$$= \frac{(x_1y_2-x_2y_1)^2}{v^2w^2}.$$

Since a quantity representing area must be positive,

$$\text{Area } V_1OV_2 = \tfrac{1}{2}|vw\sin\theta|$$
$$= \tfrac{1}{2}|x_1y_2-x_2y_1|.$$

Exercise 2a

1. Which of the following statements are meaningless?

(a) $\mathbf{v}\cdot\mathbf{w}+7 = 3\mathbf{v}$ (c) $v^2+7\mathbf{v}\cdot\mathbf{v} = 3v^2$
(b) $\mathbf{v}\cdot\mathbf{w}+7 = \mathbf{0}$ (d) $\mathbf{v}\cdot\mathbf{w}+7v^2 = 0$.

2. $ABCD$ is a kite, with $AB = BC$, $AD = DC$. Show that $\overline{AB}\cdot\overline{DA} = \overline{CB}\cdot\overline{DC}$. Deduce that the diagonals of a kite are perpendicular.

3. $ABCD$ is a parallelogram. $\overline{AB} = \mathbf{a}$, $\overline{AD} = \mathbf{b}$, $\overline{AC} = \mathbf{c}$, $\overline{BD} = \mathbf{d}$. Express $\mathbf{c}\cdot\mathbf{c}$ and $\mathbf{d}\cdot\mathbf{d}$ in terms of $\mathbf{a}$ and $\mathbf{b}$. Deduce that the sum of the squares of the four sides of a parallelogram is equal to the sum of the squares of the diagonals.

4. Give unit vectors perpendicular to (a) $\begin{pmatrix}3\\-4\end{pmatrix}$, (b) $\begin{pmatrix}0\\5\end{pmatrix}$, (c) the line joining the *points* with coordinates $(3, -4)$, $(0, 5)$.

5. $\overline{OP} = \begin{pmatrix}5\\0\end{pmatrix}$, $\overline{OQ} = \begin{pmatrix}4\\3\end{pmatrix}$. M is the mid-point of PQ. Calculate $\overline{OM}\cdot\overline{PQ}$. Verify your result by investigating what type of triangle OPQ is.

6. $\overline{OA} = \begin{pmatrix}-3\\1\end{pmatrix}$, $\overline{OB} = \begin{pmatrix}2\\1\end{pmatrix}$. Calculate angle AOB. (Do not look up its value in tables, but just find its cosine.) Derive a value for sin AOB, and use the formula $\tfrac{1}{2}ab\sin C$ to find the area of the triangle AOB. Then use the formula obtained in Example (xi) to check your answer. Finally, look for something special about the shape of triangle AOB to check your answer again.

Position Vectors

We have noted that the vector **v** has infinitely many representations that specify it in size and direction, but not position. One of these representations is $\overline{OV}$, where O is a fixed point we shall call the origin. $\overline{OV} = \mathbf{v}$ is called the position vector of V referred to O.

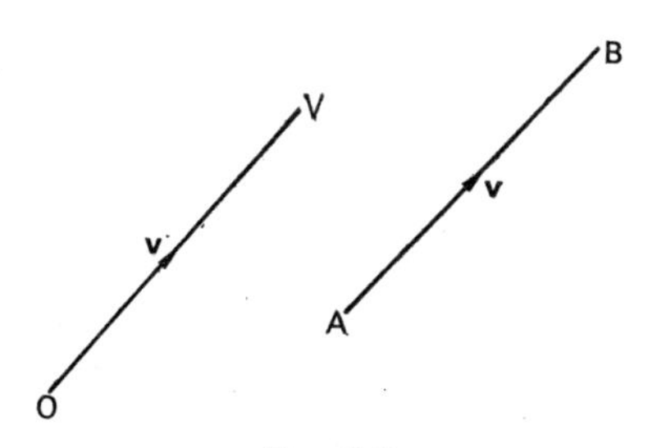

FIG. 2.6

In Fig. 2.6 $\mathbf{v} = \overline{AB} = \overline{OV}$. **v** is the position vector of the point V referred to O, but not the position vector of A or B.

We now prove a theorem concerning position vectors.

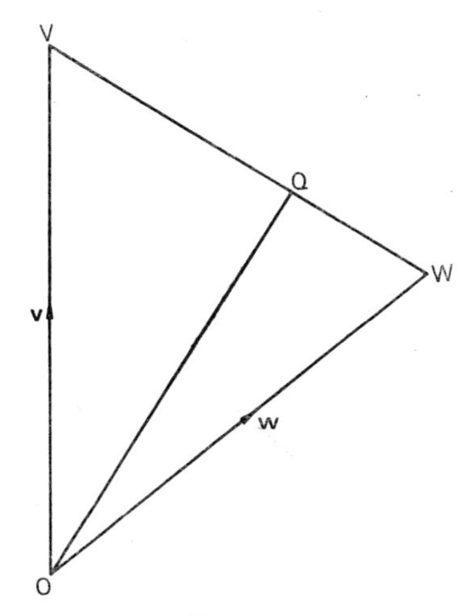

FIG. 2.7

THEOREM. *If V, W have position vectors* $\mathbf{v}$, $\mathbf{w}$ *referred to O, and Q divides VW in the ratio $l:m$, where $l+m=1$, then* $\overline{OQ} = m\mathbf{v}+l\mathbf{w}$.

Proof. From Fig. 2.7,

$$\begin{aligned}\overline{VW} &= \mathbf{w}-\mathbf{v}\\ \overline{VQ} &= l\,\overline{VW}\\ &= l(\mathbf{w}-\mathbf{v})\\ \overline{OQ} &= \overline{OV}+\overline{VQ}\\ &= \mathbf{v}+l(\mathbf{w}-\mathbf{v})\\ &= (1-l)\mathbf{v}+l\mathbf{w}\\ &= m\mathbf{v}+l\mathbf{w}.\end{aligned}$$

EXAMPLE (xii). The position vectors of points A, B, C are $\mathbf{a}$, $\mathbf{b}$, $\mathbf{c}$. AD is a median of the triangle ABC and G divides AD in the ratio $\frac{2}{3}:\frac{1}{3}$. Find the position vector of G in terms of $\mathbf{a}$, $\mathbf{b}$ and $\mathbf{c}$. Deduce a well-known theorem in geometry.

Figure 2.8 shows the triangle with an origin O. If the position vectors of D, G are $\mathbf{d}$, $\mathbf{g}$, then

$$\mathbf{d} = \tfrac{1}{2}\mathbf{b}+\tfrac{1}{2}\mathbf{c}$$

and

$$\begin{aligned}\mathbf{g} &= \tfrac{1}{3}\mathbf{a}+\tfrac{2}{3}\mathbf{d}, \quad \text{by the previous theorem,}\\ &= \tfrac{1}{3}\mathbf{a}+\tfrac{2}{3}(\tfrac{1}{2}\mathbf{b}+\tfrac{1}{2}\mathbf{c})\\ &= \tfrac{1}{3}(\mathbf{a}+\mathbf{b}+\mathbf{c}).\end{aligned}$$

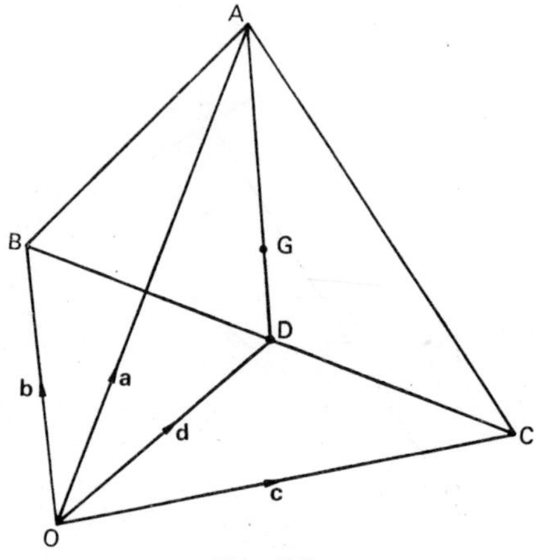

FIG. 2.8

By the symmetry of this result, we see that G must also trisect the other medians of the triangle. We deduce that the medians of the triangle are concurrent and trisect each other.

Note. G is called the *centroid* of the triangle ABC.

The Vector Equation of a Line

You are probably familiar with the coordinate equation of a straight line:

$$y = mx + c$$

where m is the gradient of the line and c its intercept on the y-axis. (A brief consultation with any elementary textbook on coordinate geometry will remind you how this equation is derived.) We shall now derive a vector equation of a straight line, and show how the coordinate equation may be derived from it.

Suppose V is a variable point on a straight line which is parallel to a given vector **b** and passes through a fixed point A, whose position vector referred to an origin O is **a**. An equation giving the position vector of V in terms of **a** and **b** will be the vector equation of the line.

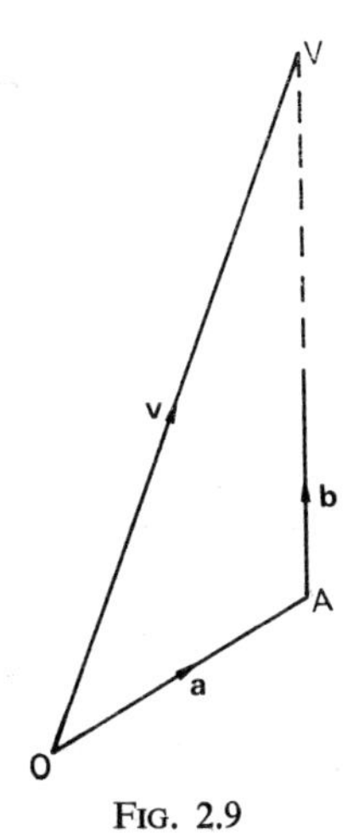

FIG. 2.9

But if V is a variable point, its position vector must vary. As is clear from Fig. 2.9, the vector $\overline{AV}$ varies, but is always parallel to $\mathbf{b}$. That is, $\overline{AV} = r\mathbf{b}$, where r is a variable real number. Hence, if $\overline{OV} = \mathbf{v}$,

$$\mathbf{v} = \mathbf{a}+r\mathbf{b}.$$

The variable r is called a *parameter*, and the above equation the *vector equation* of the line.

If $\mathbf{a}$, $\mathbf{b}$ are given as coordinate vectors, say $\mathbf{a} = \begin{pmatrix} a \\ b \end{pmatrix}$ and $\mathbf{b} = \begin{pmatrix} l \\ m \end{pmatrix}$, and the variable $\mathbf{v} = \begin{pmatrix} x \\ y \end{pmatrix}$, then we may derive from the vector equation a pair of coordinate equations:

$$x = a+lr,$$
$$y = b+mr.$$

Eliminating the parameter r from this pair of equations leads us to a single linear coordinate equation:

$$\frac{x-a}{l} = \frac{y-b}{m},$$

which is a rearrangement of the equation of the line in the form mentioned at the beginning of this section.

EXAMPLE (xiii). In Fig. 2.10 find the equations of the lines through the point (4, 1) parallel to (a) $\mathbf{v}_1$, (b) $\mathbf{v}_2$.

(a) Taking $\mathbf{a} = \begin{pmatrix} 4 \\ 1 \end{pmatrix}$, $\mathbf{v} = \begin{pmatrix} x \\ y \end{pmatrix}$, the vector equation of this line is

$$\mathbf{v} = \mathbf{a}+r\mathbf{v}_1$$

where r is a parameter. This implies

$$x = 4+3r,$$
$$y = 1+3r.$$

Eliminating r between these gives the single linear equation

$$y = x-3.$$

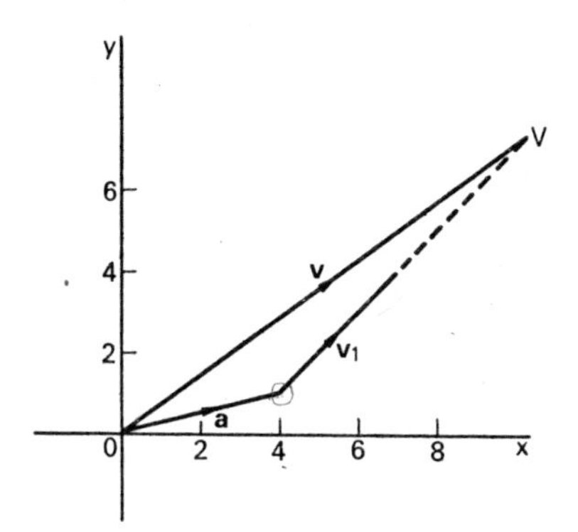

FIG. 2.10

(b) The vector equation is $\mathbf{v} = \mathbf{a}+r\mathbf{v}_2$.

This implies

$$x = 4-2r,$$
$$y = 1+0r.$$

This leads to the single equation $y = 1$.

EXAMPLE (xiv). Find the equation of the line through the points (2, 1), (3, −1).

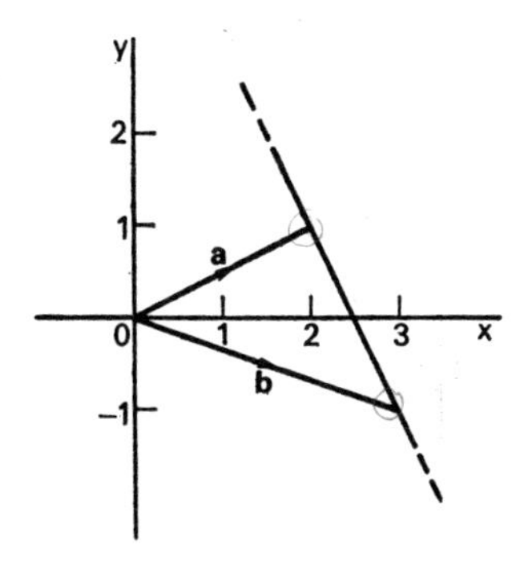

FIG. 2.11

Let $\mathbf{a} = \begin{pmatrix}2\\1\end{pmatrix}$, $\mathbf{b} = \begin{pmatrix}3\\-1\end{pmatrix}$. From Fig. 2.11 it is clear that the required line is parallel to $\mathbf{b}-\mathbf{a}$.

Its vector equation is thus $\mathbf{v} = \mathbf{a}+r(\mathbf{b}-\mathbf{a})$.

This implies $x = 2+r,$

$y = 1-2r.$

This leads to the single equation $y = -2x+5.$

EXAMPLE (xv). Find the angle α between the lines $y = 2x$, $y = 3x$.

We select any point on $y = 2x$, say (1, 2), so that the vector $\mathbf{a} = \begin{pmatrix}1\\2\end{pmatrix}$ is parallel to the line $y = 2x$.

Similarly, the vector $\mathbf{b} = \begin{pmatrix}1\\3\end{pmatrix}$ is parallel to the line $y = 3x$.

Now
$$\mathbf{a}\cdot\mathbf{b} = 1\cdot1+2\cdot3 = 7.$$

Also
$$\mathbf{a}\cdot\mathbf{b} = ab\cos\alpha = \sqrt{5}\cdot\sqrt{10}\cos\alpha.$$

Thus
$$\cos\alpha = \frac{7}{\sqrt{50}}.$$

EXAMPLE (xvi). Find the shortest distance from the point P (3, 1) to the line $4y = 3x$, and the projection of OP onto that line.

Referring to Fig. 2.12, we see that the shortest distance is

$$PN = OP\sin\alpha$$

where α is the angle between OP and the line $4y = 3x$.

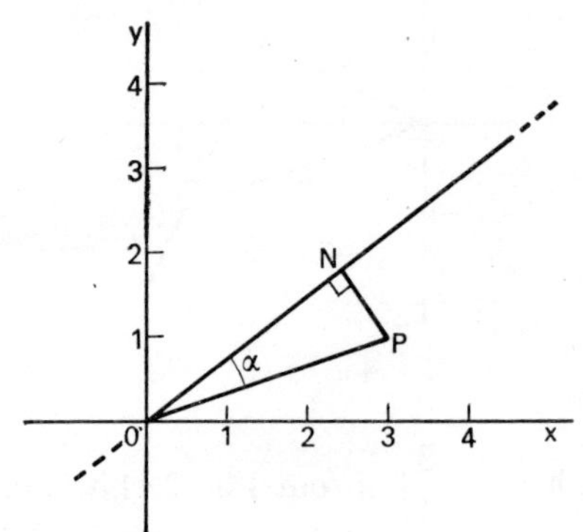

FIG. 2.12

A method similar to that used in Example (xv) gives

$$\cos\alpha = \frac{3}{\sqrt{10}}.$$

Thus $$\sin\alpha = \frac{1}{\sqrt{10}}.$$

Therefore the required shortest distance is $\sqrt{10}\cdot\frac{1}{\sqrt{10}} = 1.$

The projection of OP onto the line is $ON = OP\cos\alpha$
$= 3.$

The Distributive Property of Inner Product

Finally we show a proof of the distributive property that was illustrated in Example (vi) and used in much of the subsequent work in this chapter. The other properties of vectors which we have been content to illustrate rather than to prove in these two chapters are in fact even simpler to prove than this property.

With reference to Fig. 2.13, we shall prove that

$$\mathbf{w}\cdot(\mathbf{u}+\mathbf{v}) = \mathbf{w}\cdot\mathbf{u}+\mathbf{w}\cdot\mathbf{v}.$$

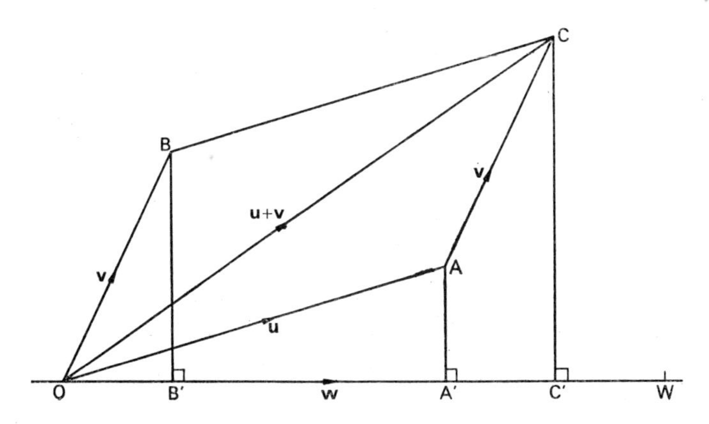

FIG. 2.13

In Fig. 2.13,

$$\overline{OA} = \mathbf{u}, \quad \overline{OB} = \mathbf{v}, \quad \overline{OC} = \mathbf{u}+\mathbf{v}, \quad \overline{OW} = \mathbf{w}.$$

$$\begin{aligned} \mathbf{w}\cdot(\mathbf{u}+\mathbf{v}) &= w\cdot OC \cos COC' \\ &= w\cdot OC'. \\ \mathbf{w}\cdot\mathbf{u}+\mathbf{w}\cdot\mathbf{v} &= w\cdot OA \cos AOA' + w\cdot OB \cos BOB' \\ &= w\cdot OA' + w\cdot OB' \\ &= w\cdot OC'. \end{aligned}$$

Thus $$\mathbf{w}\cdot(\mathbf{u}+\mathbf{v}) = \mathbf{w}\cdot\mathbf{u}+\mathbf{w}\cdot\mathbf{v}.$$

Note. If in the above property we put $\mathbf{w} = \mathbf{s}+\mathbf{t}$, we have

$$(\mathbf{s}+\mathbf{t})\cdot(\mathbf{u}+\mathbf{v}) = (\mathbf{s}+\mathbf{t})\cdot\mathbf{u}+(\mathbf{s}+\mathbf{t})\cdot\mathbf{v}.$$

But $$(\mathbf{s}+\mathbf{t})\cdot\mathbf{u} = \mathbf{s}\cdot\mathbf{u}+\mathbf{t}\cdot\mathbf{u}.$$

Thus $$(\mathbf{s}+\mathbf{t})\cdot(\mathbf{u}+\mathbf{v}) = \mathbf{s}\cdot\mathbf{u}+\mathbf{t}\cdot\mathbf{u}+\mathbf{s}\cdot\mathbf{v}+\mathbf{t}\cdot\mathbf{v}.$$

This justifies what we have been doing, manipulating brackets with techniques similar to those we use in the algebra of numbers.

Summary of Chapter 2

We have developed some algebra of vectors and used it to solve some problems in geometry.

We have defined the inner product of two vectors and seen that this operation has a commutative and a distributive property. A rule followed for the inner product of coordinate vectors, namely:

$$\begin{pmatrix} x_1 \\ y_1 \end{pmatrix}\cdot\begin{pmatrix} x_2 \\ y_2 \end{pmatrix} = x_1x_2+y_1y_2.$$

We have introduced unit vectors and position vectors, and we proved a theorem concerning the position vector of a point dividing a line segment in a particular ratio.

We have established the vector equation of a line referred to an origin O.

Exercise 2b

1. Look through Chapters 1 and 2 for the properties of vectors that we have illustrated and noted without actually proving in general. Select two and prove them generally true for any vectors, using some elementary geometry of similar or congruent triangles.

2. *ABC* is a triangle. Its altitudes *AD*, *BE* meet at *H*. The position vectors of *A*, *B*, *C* referred to *H* as origin are **a**, **b**, **c**. Derive two vector equations from the facts that *AH* and *BH* are perpendicular to *BC*, *AC* respectively. Deduce from these equations a third equation implying that *CH* is perpendicular to *AB*. What geometry theorem have you proved?

3. *ABC* is a triangle. The perpendicular bisectors of *BC*, *AC* meet at *O*. Take *O* as origin and use a method of position vectors similar to that of question 2 to deduce that the line joining *O* to the mid-point of *AB* is perpendicular to *AB*. What geometry theorem have you proved?

4. Find (i) the vector equation, (ii) the coordinate equation of the following lines:

(a) parallel to the vector $\mathbf{s} = \begin{pmatrix} 0 \\ 1 \end{pmatrix}$, passing through the point (2, 3);

(b) parallel to the vector $\mathbf{t} = \begin{pmatrix} 3 \\ -2 \end{pmatrix}$, passing through the point (2, 3);

(c) passing through the points (2, 3), (0, 1).

5. Find the angles between the lines of question 4.

6. Find the perpendicular distance of the point P (4, 3) from the line $y = 2x$, and find the projection of OP onto the line.

CHAPTER 3

VECTORS IN THREE DIMENSIONS

Introduction

You may have noticed that although the coordinate vectors we have used so far refer to plane geometry, the definitions of Chapters 1 and 2 need not be confined to plane geometry, but apply in three-dimensional space. Imagine, or construct with pencils or knitting needles, three vectors $\mathbf{v}_1$, $\mathbf{v}_2$, $\mathbf{v}_3$ which are not coplanar, and then find a vector which is $\mathbf{v}_1+\mathbf{v}_2+\mathbf{v}_3$.

Points in space may be referred to an origin O by position vectors, as the following example will illustrate.

EXAMPLE (i). Prove that the joins of the mid-points of opposite edges of a tetrahedron bisect each other.

A tetrahedron is the shape formed by joining four points in space with six straight edges. (The edges do not necessarily have equal lengths.) Figure 3.1 is an illustration of such a three-dimensional figure, in which the edge BD is behind AC. X, Y are the mid-points of the opposite edges AD, BC. O is some origin, referred to which the position vectors of A, B, C, D, X, Y are $\mathbf{a}$, $\mathbf{b}$, $\mathbf{c}$, $\mathbf{d}$, $\mathbf{x}$, $\mathbf{y}$.

Now $\mathbf{x} = \frac{1}{2}\mathbf{a}+\frac{1}{2}\mathbf{d}$, and $\mathbf{y} = \frac{1}{2}\mathbf{b}+\frac{1}{2}\mathbf{c}$.

The mid-point of XY will be P, whose position vector, $\mathbf{p}$, satisfies:

$$\begin{aligned}\mathbf{p} &= \tfrac{1}{2}\mathbf{x}+\tfrac{1}{2}\mathbf{y}\\ &= \tfrac{1}{4}(\mathbf{a}+\mathbf{b}+\mathbf{c}+\mathbf{d}).\end{aligned}$$

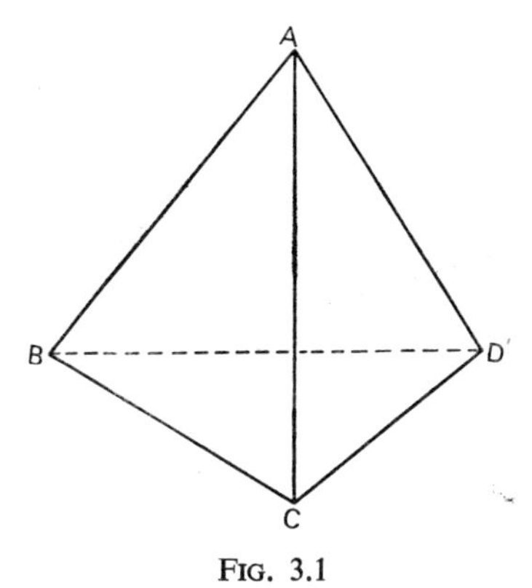

FIG. 3.1

From the symmetry of this result, we see that the point P must bisect the joins of the mid-points of the other two pairs of opposite edges of the tetrahedron.

What is the analogy of this result in two dimensions (i.e. if A, B, C, D are coplanar points)?

Three-dimensional Coordinate Vectors

We now introduce a coordinate system which will enable us to use coordinate vectors applying to three-dimensional space.

Notation. To define the position of a point in three-dimensional space, it is convenient to introduce three mutually perpendicular axes of reference, Ox, Oy, Oz. Figure 3.2 illustrates a right-handed system of such axes. Imagining that Oy points in some direction through the book, and Ox points out of it, then a right-handed screw rotating in the sense from Ox to Oy would penetrate in the direction of Oz.

The coordinates (x, y, z) define the position of a point V whose distance from the y–z-plane is x, from the x–z-plane is y, and from the x–y-plane is z.

The position vector of V referred to O is $\overline{OV}$, and we use the nota-

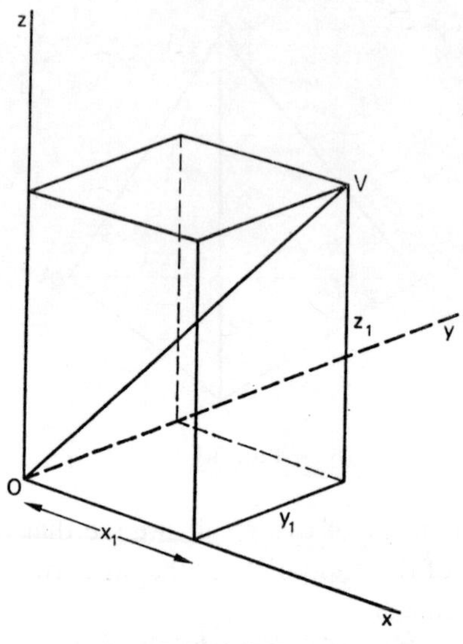

FIG. 3.2

tion $\overline{OV} = \begin{pmatrix} x \\ y \\ z \end{pmatrix}$. $\mathbf{v} = \begin{pmatrix} x \\ y \\ z \end{pmatrix}$ is any vector whose size and direction are the same as those of $\overline{OV}$.

The unit vectors $\begin{pmatrix} 1 \\ 0 \\ 0 \end{pmatrix}, \begin{pmatrix} 0 \\ 1 \\ 0 \end{pmatrix}, \begin{pmatrix} 0 \\ 0 \\ 1 \end{pmatrix}$ are denoted by **i**, **j**, **k**.

Rules for Three-dimensional Coordinate Vectors

The following rules may be shown to hold, using methods of proof similar to those used in the two-dimensional cases.

If $\mathbf{v} = \begin{pmatrix} x_1 \\ y_1 \\ z_1 \end{pmatrix}$, $\mathbf{w} = \begin{pmatrix} x_2 \\ y_2 \\ z_2 \end{pmatrix}$, then

1. $\mathbf{v}+\mathbf{w} = \begin{pmatrix} x_1+x_2 \\ y_1+y_2 \\ z_1+z_2 \end{pmatrix}$,

2. $k\mathbf{v} = \begin{pmatrix} kx_1 \\ ky_1 \\ kz_1 \end{pmatrix}$,

3. $\mathbf{v}\cdot\mathbf{w} = x_1x_2+y_1y_2+z_1z_2$,

and in particular $\mathbf{v}\cdot\mathbf{v} = x_1^2+y_1^2+z_1^2$.

EXAMPLE (ii). Give vectors (a) parallel to, (b) perpendicular to $\mathbf{v} = \begin{pmatrix} 2 \\ 3 \\ 0 \end{pmatrix}$.

(a) Since, by Definition 1.4, $k\mathbf{v}$ is a vector parallel to $\mathbf{v}$, suitable vectors include $\begin{pmatrix} 4 \\ 6 \\ 0 \end{pmatrix}$, $\begin{pmatrix} -6 \\ -9 \\ 0 \end{pmatrix}$, etc.

(b) By Rule 3, the vector $\mathbf{w} = \begin{pmatrix} x \\ y \\ z \end{pmatrix}$ is perpendicular to the given vector provided

$$2x+3y+0z = 0.$$

Thus suitable vectors include $\begin{pmatrix} 3 \\ -2 \\ 0 \end{pmatrix}$, $\begin{pmatrix} -6 \\ 4 \\ 5 \end{pmatrix}$, $\begin{pmatrix} 0 \\ 0 \\ 7 \end{pmatrix}$, etc.

Try to imagine the directions of the vectors mentioned in this example. You will find that it is seldom very illuminating to sketch them on a two-dimensional page. However, we illustrate the vector $\mathbf{v}$ of this example in Fig. 3.3.

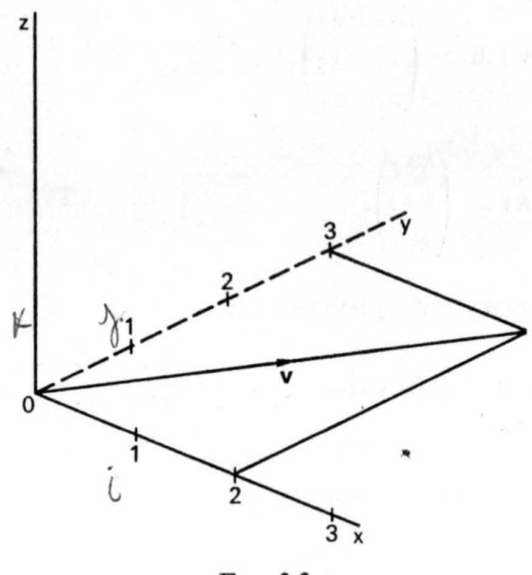

FIG. 3.3

EXAMPLE (iii). Find the size of the vector $\mathbf{v} = \begin{pmatrix} 2 \\ 3 \\ 0 \end{pmatrix}$, and the angles α, β it makes with Ox and Oy.

By Rule 3, the size of the vector is v, where

$$v^2 = 2^2+3^2+0^2.$$

Thus $$v = \sqrt{13}.$$

By Definition 2.2, $$\mathbf{v}\cdot\mathbf{i} = \sqrt{13}\cdot 1 \cos\alpha.$$

But by Rule 3, $$\mathbf{v}\cdot\mathbf{i} = 2\cdot 1+3\cdot 0+0\cdot 0.$$

Thus $$\cos\alpha = \frac{2}{\sqrt{13}}.$$

Similarly, $$\mathbf{v}\cdot\mathbf{j} = \sqrt{13}\cos\beta = 3.$$

Thus $$\cos\beta = \frac{3}{\sqrt{13}}.$$

Can you see how these results confirm the fact that $\alpha+\beta = 90°$?

EXAMPLE (iv). P is the point $(2, 3, 0)$, Q is $(2, 4, -1)$, R is $(1, 3, -1)$. Find $\overline{QP}$, $\overline{QR}$, angle PQR, and the area of triangle PQR.

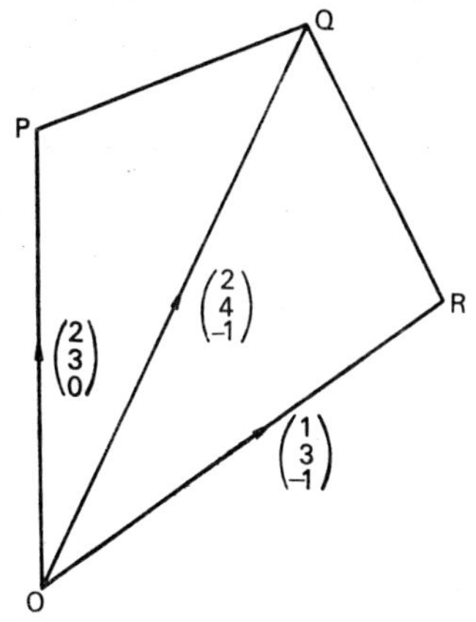

FIG. 3.4

$$\overline{QP} = \overline{OP} - \overline{OQ} \qquad\qquad \overline{QR} = \overline{OR} - \overline{OQ}$$

$$= \begin{pmatrix} 2 \\ 3 \\ 0 \end{pmatrix} - \begin{pmatrix} 2 \\ 4 \\ -1 \end{pmatrix} \qquad\qquad = \begin{pmatrix} -1 \\ -1 \\ 0 \end{pmatrix}.$$

$$= \begin{pmatrix} 0 \\ -1 \\ 1 \end{pmatrix}.$$

$$QP = QR = \sqrt{2}.$$

Thus $\quad \overline{QP} \cdot \overline{QR} = \sqrt{2} \cdot \sqrt{2} \cos PQR.$

But $\quad \overline{QP} \cdot \overline{QR} = 0 \cdot -1 + -1 \cdot -1 + 1 \cdot 0.$

Thus $\quad \cos PQR = \frac{1}{2}.$

$$\text{Area of triangle } PQR = \tfrac{1}{2} PQ \cdot QR \sin PQR$$
$$= \sqrt{2} \cdot \sqrt{2} \cdot \frac{\sqrt{3}}{2}$$
$$= \sqrt{3}.$$

Can you deduce that triangle PQR is equilateral?

The Vector Product

Many geometric problems involve finding a vector which is perpendicular to each of two given vectors. (The vector **k**, for instance, is perpendicular to **i** and **j**, as are 2**k** and −3**k**.) In Definition 3.1 we shall define a vector product of two vectors as a third vector which is perpendicular to them both. Like the inner product, the vector product may at first seem a rather arbitrary quantity, but it is in fact another useful tool.

DEFINITION 3.1. *The vector product of two vectors* **v**, **w** *is the vector* $(vw \sin \alpha)\mathbf{u}$, *where* α *is the angle between* **v** *and* **w**, *and* **u** *is a unit vector perpendicular to each of* **v** *and* **w** *so that* **v**, **w** *and* **u** *form a right-handed system.* (That is, a screw rotating in the sense from **v** to **w** would penetrate in the direction of **u**. In Fig. 3.5, imagine **v** and **w** in the plane of the page and **u** penetrating it.)

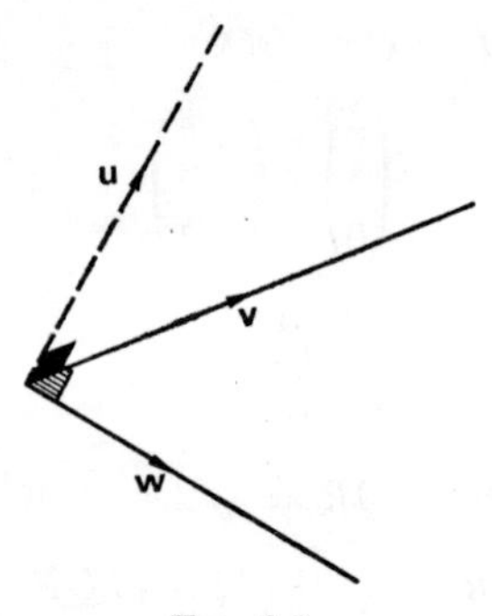

FIG. 3.5

Notation. The symbol for vector product is ×, so that

$$\mathbf{v} \times \mathbf{w} = (vw \sin \alpha)\mathbf{u}.$$

EXAMPLE (v). Calculate $\mathbf{i}\times\mathbf{i}$, $\mathbf{i}\times\mathbf{j}$, $\mathbf{j}\times\mathbf{i}$.

$$\mathbf{i}\times\mathbf{i} = \mathbf{0},$$

since $\sin 0° = 0$.

The vectors **i**, **j** and **k** form a right-handed system, so that

$$\mathbf{i}\times\mathbf{j} = (1\cdot 1 \sin 90°)\mathbf{k} = \mathbf{k}.$$

The vectors **j**, **i** and $-\mathbf{k}$ form a right-handed system, so that

$$\mathbf{j}\times\mathbf{i} = -\mathbf{k}.$$

Example (v) illustrates that the new operation of vector product is not commutative. We may generalise the result of the example, stating:

$$\begin{aligned} \mathbf{i}\times\mathbf{i} &= \mathbf{j}\times\mathbf{j} = \mathbf{k}\times\mathbf{k} = \mathbf{0}, \\ \mathbf{i}\times\mathbf{j} &= -(\mathbf{j}\times\mathbf{i}) = \mathbf{k}, \\ \mathbf{j}\times\mathbf{k} &= -(\mathbf{k}\times\mathbf{j}) = \mathbf{i}, \\ \mathbf{k}\times\mathbf{i} &= -(\mathbf{i}\times\mathbf{k}) = \mathbf{j}. \end{aligned}$$

Can you see a connection between the cyclic order of the letters **i**, **j**, **k** in the above and the sign of the vector product?

EXAMPLE (vi). Calculate (a) $\mathbf{v}\times\mathbf{i}$, (b) $\mathbf{v}\times\mathbf{j}$, (c) $\mathbf{v}\times(\mathbf{i}+\mathbf{j})$, where $\mathbf{v} = \begin{pmatrix} 2 \\ 3 \\ 0 \end{pmatrix}$.

(a) We require a unit vector perpendicular to **v** and to **i**. Any vector perpendicular to **i** is of the form $\begin{pmatrix} 0 \\ a \\ b \end{pmatrix}$, where a and b are real numbers.

In Example (ii) we gave a selection of vectors perpendicular to **v**. Of them, those of the form $\begin{pmatrix} 0 \\ 0 \\ c \end{pmatrix}$ are also perpendicular to **i**. The unit vector of this form forming a right-handed system with **v** and **i** is $-\mathbf{k}$.

From Example (ii), we have $v = \sqrt{13}$ and $\sin\alpha = \dfrac{3}{\sqrt{13}}$, where α is the angle between **v** and **i**.

Thus

$$\mathbf{v}\times\mathbf{i} = \sqrt{13}\cdot 1\cdot\frac{3}{\sqrt{13}}(-\mathbf{k})$$

$$= -3\mathbf{k}.$$

(b) The vectors $\mathbf{v}$, $\mathbf{j}$ and $\mathbf{k}$ form a right-handed system. With reference again to Example (ii),

$$\mathbf{v}\times\mathbf{j} = \sqrt{13}\cdot 1\cdot\frac{2}{\sqrt{13}}\cdot\mathbf{k},$$

$$= 2\mathbf{k}.$$

(c) The vectors $\mathbf{v}$, $(\mathbf{i}+\mathbf{j})$ and $-\mathbf{k}$ form a right-handed system. The angle between $\mathbf{v}$ and $(\mathbf{i}+\mathbf{j})$ is γ, where

$$\mathbf{v}\cdot(\mathbf{i}+\mathbf{j}) = \sqrt{13}\cdot\sqrt{2}\cos\gamma$$

and $$\mathbf{v}\cdot(\mathbf{i}+\mathbf{j}) = 2\cdot 1+3\cdot 1+0\cdot 0 = 5.$$

Thus $$\cos\gamma = \frac{5}{\sqrt{26}}, \quad \text{and} \quad \sin\gamma = \frac{1}{\sqrt{26}}.$$

Thus $$\mathbf{v}\times(\mathbf{i}+\mathbf{j}) = \sqrt{13}\cdot\sqrt{2}\cdot\frac{1}{\sqrt{26}}(-\mathbf{k})$$

$$= -\mathbf{k}.$$

Example (vi) illustrates the distributive property:

$$\mathbf{u}\times(\mathbf{v}+\mathbf{w}) = (\mathbf{u}\times\mathbf{v})+(\mathbf{u}\times\mathbf{w}).$$

We shall not prove this property in this book, but refer the interested reader to Macbeath, *Elementary Vector Algebra.*

Note. As with other product signs, we use the convention

$$\mathbf{u}\times\mathbf{v}+\mathbf{u}\times\mathbf{w} \quad \text{to mean} \quad (\mathbf{u}\times\mathbf{v})+(\mathbf{u}\times\mathbf{w}).$$

Rule for Vector Product of Coordinate Vectors

Taking

$$\mathbf{v} = \begin{pmatrix} x_1 \\ y_1 \\ z_1 \end{pmatrix}, \quad \mathbf{w} = \begin{pmatrix} x_2 \\ y_2 \\ z_2 \end{pmatrix},$$

we shall express $\mathbf{v}\times\mathbf{w}$ in terms of x_1, y_1, z_1, x_2, y_2, z_2. Assuming the distributive property which we illustrated in the last example, then

$$\begin{aligned}\mathbf{v}\times\mathbf{w} &= (x_1\mathbf{i}+y_1\mathbf{j}+z_1\mathbf{k})\times(x_2\mathbf{i}+y_2\mathbf{j}+z_2\mathbf{k})\\ &= x_1x_2\mathbf{i}\times\mathbf{i}+x_1y_2\mathbf{i}\times\mathbf{j}+x_1z_2\mathbf{i}\times\mathbf{k}+y_1x_2\mathbf{j}\times\mathbf{i}+y_1y_2\mathbf{j}\times\mathbf{j}\\ &\quad +y_1z_2\mathbf{j}\times\mathbf{k}+z_1x_2\mathbf{k}\times\mathbf{i}+z_1y_2\mathbf{k}\times\mathbf{j}+z_1z_2\mathbf{k}\times\mathbf{k}\\ &= (y_1z_2-z_1y_2)\,\mathbf{i}+(z_1x_2-x_1z_2)\,\mathbf{j}+(x_1y_2-y_1x_2)\mathbf{k}.\end{aligned}$$

Check Example (vi), using this rule for the vector product of co-ordinate vectors.

EXAMPLE (vii). Calculate $\overline{QP}\times\overline{QR}$, where P, Q and R are the points specified in Example (iv). Deduce the area of triangle PQR.

From Example (iv), $\overline{QP} = \begin{pmatrix}0\\-1\\1\end{pmatrix}$, $\overline{QR} = \begin{pmatrix}-1\\-1\\0\end{pmatrix}$.

Thus $\qquad \overline{QP}\times\overline{QR} = -\mathbf{i}-\mathbf{j}+\mathbf{k}.$

Now

$$\begin{aligned}\text{Area of triangle } PQR &= \tfrac{1}{2}QP\cdot QR\sin PQR\\ &= \tfrac{1}{2}\text{ the size of the vector } \overline{QP}\times\overline{QR}\\ &= \frac{\sqrt{3}}{2}.\end{aligned}$$

Points, Lines and Planes

In space, any two points are collinear—that is, a straight line may be drawn passing through both of them. But a third point may or may not be collinear with the other two. How can we test whether three points are collinear?

EXAMPLE (viii). Is the point S (2, 2, 1) collinear with the points P, Q of Example (iv)?

If the vectors $\overline{QP}$, $\overline{QS}$ are in the same direction, then it follows that the points are collinear.

$$\overline{QP} = \begin{pmatrix}0\\-1\\1\end{pmatrix}, \quad\text{and}\quad \overline{QS} = \overline{OS}-\overline{OQ} = \begin{pmatrix}0\\-2\\2\end{pmatrix}.$$

Thus $\overline{QS}$ is a scalar multiple of $\overline{QP}$, implying that $\overline{QS}$ and $\overline{QP}$ have the same direction.

Thus P, Q, S are collinear.

Any three points in space are coplanar—that is, a plane may be constructed passing through all of them. But a fourth point may or may not be coplanar with the other three. How can we test whether four points are coplanar?

EXAMPLE (ix). Is the point T (1, 2, 0) coplanar with the points P, Q, R of Example (iv)?

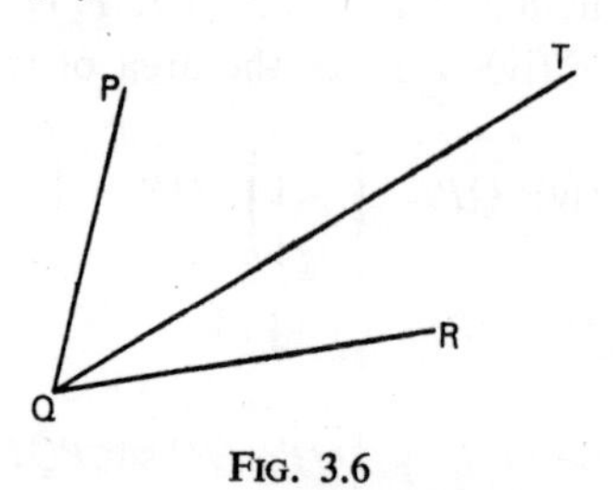

FIG. 3.6

If P, Q, R, T are coplanar, then

$$\text{angle } PQT \pm \text{angle } TQR = \text{angle } PQR.$$

If this equality does not hold, then the lines PQ, TQ, RQ will not be coplanar. Satisfy yourself of this by observing a model such as the corner of a box, or by representing the three lines by three pencils.

$$\text{Now} \quad \overline{QP} = \begin{pmatrix} 0 \\ -1 \\ 1 \end{pmatrix}, \quad \overline{QT} = \begin{pmatrix} -1 \\ -2 \\ 1 \end{pmatrix}, \quad \overline{QR} = \begin{pmatrix} -1 \\ -1 \\ 0 \end{pmatrix}.$$

Thus $\qquad \overline{QT}\cdot\overline{QP} = 3 = \sqrt{2}\cdot\sqrt{6}\cos PQT,$

and $\qquad \cos PQT = \dfrac{\sqrt{3}}{2}.$

Also $\qquad \overline{QT}\cdot\overline{QR} = 3 = \sqrt{2}\cdot\sqrt{6}\cos TQR,$

so $\qquad \cos TQR = \dfrac{\sqrt{3}}{2}.$

In Example (iv) we saw that $PQR = 60°$. Thus

$$PQT+TQR = PQR,$$

and the four points are coplanar.

Exercise 3a

1. $ABCD$ is a tetrahedron. G_1 is the centroid of triangle BCD, and G_2, G_3, G_4 are similarly defined. By considering position vectors, show that AG_1, BG_2, CG_3, DG_4 are concurrent at a point G which divides each of them in the ratio 1 : 3.
2. P is the point $(-3, 2, 1)$. Find $\overline{OP}$, OP and the angle between OP and Ox. Hence find the area of the triangle OPQ, where Q is the point (1, 0, 0).
3. A is the point (1, 2, 3), B (2, 0, 4), C (1, -2, 1). Find AC, AB, and angle BAC. Hence find the area of triangle ABC.
4. With reference to question 2, calculate $\overline{OP}\times\overline{OQ}$ and deduce the area of triangle OPQ, checking your previous result.
5. With reference to question 3, use a method involving vector product to calculate the area of triangle ABC.
6. Test whether the points A, B of question 3 and P of question 2 are collinear.
7. Test whether the points A, B, C of question 3 are coplanar with O. Then look for something special about the shape of $OABC$ and use your discovery to check your results for this question and question 3.

The Equations of a Line

In Chapter 2 we obtained the vector equation of a line:

$$\mathbf{v} = \mathbf{a}+r\mathbf{b},$$

where $\mathbf{v}$ gives the position vector of any point V on the line in terms of the position vector $\mathbf{a}$ of a fixed point on the line and the vector $\mathbf{b}$, parallel to the line, r being a parameter (Fig. 2.9).

If $\mathbf{a}$, $\mathbf{b}$ are given as three-dimensional coordinate vectors, such as

$$\mathbf{a} = \begin{pmatrix} a \\ b \\ c \end{pmatrix} \text{ and } \mathbf{b} = \begin{pmatrix} l \\ m \\ n \end{pmatrix}, \text{ and the variable } \mathbf{v} = \begin{pmatrix} x \\ y \\ z \end{pmatrix},$$

then we may derive from the vector equation three coordinate equations:

$$\begin{aligned} x &= a+lr, \\ y &= b+mr, \\ z &= c+nr. \end{aligned}$$

If l, m and n are all non-zero, then eliminating the parameter r from these three equations leads to a pair of linear equations in x, y and z:

$$\frac{x-a}{l} = \frac{y-b}{m} = \frac{z-c}{n},$$

called the coordinate equations of the line. If one or two of l, m, n are zero, the pair of equations take a different form. For instance, if $l = 0$, $m \neq 0$, $n \neq 0$, then the equations are

$$x = a, \quad \frac{y-b}{m} = \frac{z-c}{n}.$$

EXAMPLE (x). Find the equations of the line through the point R (1, 3, −1) parallel to (a) $\overline{OQ} = \begin{pmatrix} 1 \\ 2 \\ -1 \end{pmatrix}$, (b) Ox.

(a) The vector equation is $\mathbf{v} = \overline{OR}+r\overline{OQ}$, where r is a parameter.

This implies
$$\begin{aligned} x &= 1+1r, \\ y &= 3+2r, \\ z &= -1-r. \end{aligned}$$

Eliminating r between these gives the pair of equations:

$$\frac{x-1}{1} = \frac{y-3}{2} = \frac{z+1}{-1}.$$

Note. We may use either the parametric equations or the coordinate equations to obtain a set of points on the line, such as (1, 3, −1), (2, 5, −2), (−1, −1, 1), etc,

(b) The vector equation is $\mathbf{v} = \overline{OR}+r\mathbf{i}$.

This implies
$$x = 1+r,$$
$$y = 3+0r,$$
$$z = -1+0r.$$

This leads to the pair of equations
$$y = 3,$$
$$z = -1.$$

Note. It is impossible to write this particular pair of equations in the form we used for part (a) of this question. Why?

EXAMPLE (xi). Find the equations of the line through P, Q of Example (iv) and find if this line meets the lines of Example (x).

The line PQ has the direction of the vector $\overline{PQ} = \begin{pmatrix} 0 \\ 1 \\ -1 \end{pmatrix}$.

Its vector equation is $\mathbf{v} = \overline{OP}+r\overline{PQ}$.

This implies
$$x = 2+0r,$$
$$y = 3+r,$$
$$z = 0-r.$$

This leads to the pair of equations
$$x = 2,$$
$$y+z = 3.$$

The line PQ meets line (a) of Example (x) at a point whose x coordinate is 2. The point $(2, 5, -2)$ satisfies the equations of both the lines, and so it is the point where the two lines meet.

If line PQ meets the line (b) of Example (x), it will be at a point whose y coordinate is 3 and whose z coordinate is -1. But such a point will not satisfy the equation $y+z = 3$. Thus these two lines do not meet; they are skew lines.

Note. We could now tackle Example (viii) a second way, by testing whether the coordinates of S satisfy the equations of PQ. Check that they do.

The Equation of a Plane

What information is necessary to specify a plane? Hold a flat book in some position to represent part of a plane. The neatest way of describing the "tilt" of the plane is to specify a vector which is *normal* to the plane—that is, perpendicular to any line that may be drawn in the plane.

Suppose that an origin O is selected and that P is a point on the plane so that OP is normal to the plane. We shall derive an equation giving $\mathbf{v}$, the position vector of any point V on the plane, in terms of $\mathbf{u}$, the unit vector in the direction of OP, and p, the length OP.

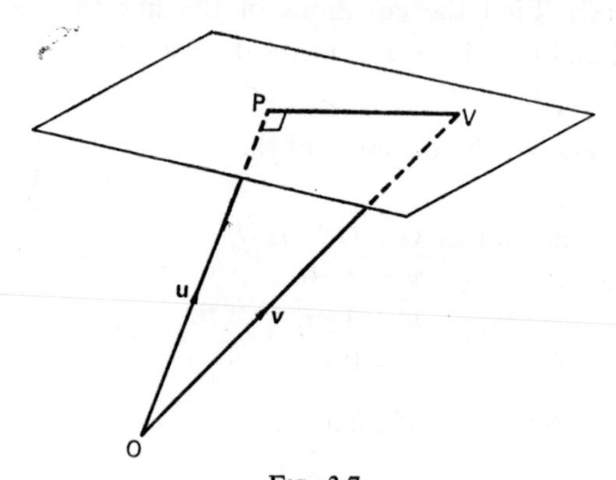

FIG. 3.7

For all positions of V, angle OPV is a right angle.

Therefore $$p = v \cos POV.$$

But $$\mathbf{v}\cdot\mathbf{u} = v \cos POV,$$

so that $$\mathbf{v}\cdot\mathbf{u} = p.$$

This is the vector equation of the plane.

If $\mathbf{u} = \begin{pmatrix} l \\ m \\ n \end{pmatrix}$ and $\mathbf{v} = \begin{pmatrix} x \\ y \\ z \end{pmatrix}$, the vector equation implies

$$lx+my+nz = p.$$

This is the coordinate equation of the plane.

Note. The equation $lx+my+nz = p$ implies also

$$klx+kmy+knz = kp, \quad \text{or} \quad \mathbf{v}\cdot k\mathbf{u} = kp.$$

The vector $\begin{pmatrix} kl \\ km \\ kn \end{pmatrix}$ is normal to the plane. So in general, the linear equation

$$ax+by+cz = d$$

represents a plane which is perpendicular to the vector $\begin{pmatrix} a \\ b \\ c \end{pmatrix}$.

EXAMPLE (xii). Find the equation of the plane through S (2, 2, 1), perpendicular to OS.

The unit vector in the direction of OS is $\mathbf{u} = \begin{pmatrix} \frac{2}{3} \\ \frac{2}{3} \\ \frac{1}{3} \end{pmatrix}$.

Thus the vector equation of the plane is

$$\mathbf{v}\cdot\mathbf{u} = 3 \qquad (OS = 3).$$

This implies the coordinate equation

$$\tfrac{2}{3}x+\tfrac{2}{3}y+\tfrac{1}{3}z = 3.$$

EXAMPLE (xiii). Find the perpendicular distance from O to the planes
(a) $x+y = 3$, (b) $x = 0$, (c) $ax+by+cz = d$.

(a) This equation is of the form $\mathbf{v}\cdot k\mathbf{u} = kp$, where $k\mathbf{u} = \begin{pmatrix} 1 \\ 1 \\ 0 \end{pmatrix}$.

Since $\mathbf{u}$ is a unit vector, $k = \sqrt{2}$, and the distance from O to the plane is $p = 3/\sqrt{2}$.

(b) This plane passes through O.

(c) This equation is of the form $\mathbf{v} \cdot k\mathbf{u} = kp$, where $k\mathbf{u} = \begin{pmatrix} a \\ b \\ c \end{pmatrix}$.

Since $\mathbf{u}$ is a unit vector, $k = \sqrt{(a^2+b^2+c^2)}$, and the distance from O to the plane is $p = \dfrac{|d|}{\sqrt{(a^2+b^2+c^2)}}$.

EXAMPLE (xiv). Find the equation of the plane through P, Q, R of Example (iv).

Suppose the equation of the plane is $ax+by+cz = d$. Then the vector $\begin{pmatrix} a \\ b \\ c \end{pmatrix}$ must be perpendicular to the plane, and thus also to

$\overline{PQ} = \begin{pmatrix} 0 \\ 1 \\ -1 \end{pmatrix}$ and $\overline{QR} = \begin{pmatrix} -1 \\ -1 \\ 0 \end{pmatrix}$.

A vector perpendicular to both PQ and QR is $\overline{PQ} \times \overline{QR} = \begin{pmatrix} 1 \\ -1 \\ -1 \end{pmatrix}$.

The coordinate equation may thus be written

$$x-y-z = d.$$

Using the fact that the coordinates of P must satisfy the equation, we deduce that $d = -1$.

The equation of the plane is therefore

$$x-y-z = -1.$$

Check that the coordinates of Q and R satisfy this equation.

Note. We could now tackle Example (ix) a second way, by testing whether the coordinates of T satisfy the equation of the plane through P, Q and R. Check that they do.

EXAMPLE (xv). Find where the plane of Example (xiv) meets the plane of Example (xiii) (a).

Any points lying on both planes have coordinates satisfying simultaneously:

$$x-y-z = -1,$$
$$x+y = 3.$$

In fact, the planes intersect in the straight line defined by this pair of equations. They may be rearranged to the more conventional form:

$$\frac{x}{1} = \frac{y-3}{-1} = \frac{z+2}{2}.$$

DEFINITION 3.2. *The angle between two planes is the angle formed between their normal vectors, chosen so that it is acute (or 90°).*

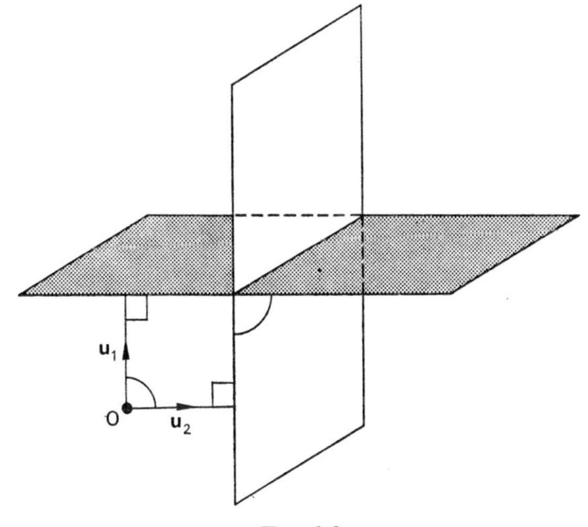

FIG. 3.8

Figure 3.8 illustrates that this angle is the same size as the angle which common sense might interpret as the angle between the planes.

EXAMPLE (xvi). Find the angle between the planes of Example (xv).

Normal vectors to the two planes are $\mathbf{a} = \begin{pmatrix} 1 \\ -1 \\ -1 \end{pmatrix}$, $\mathbf{b} = \begin{pmatrix} 1 \\ 1 \\ 0 \end{pmatrix}$.

The angle between **a** and **b** is α, where

$$\mathbf{a}\cdot\mathbf{b} = \sqrt{3}\cdot\sqrt{2}\cos\alpha$$

and also $\mathbf{a}\cdot\mathbf{b} = 1-1 = 0$.

Thus $\cos\alpha = 0$, and the planes are perpendicular.

DEFINITION 3.3. *The angle between a line and a plane is the complement of the angle between the line and the normal vector to the plane, chosen so that it is acute (or 90°).*

EXAMPLE (xvii). Find the angle between the plane $x+y = 2$ and the line

$$\frac{x-1}{1} = \frac{y}{2} = \frac{z-1}{1}.$$

A vector perpendicular to the plane is $\mathbf{a} = \begin{pmatrix}1\\1\\0\end{pmatrix}$.

A vector parallel to the line is $\mathbf{b} = \begin{pmatrix}1\\2\\1\end{pmatrix}$.

The angle between **a** and **b** is α, where

$$\mathbf{a}\cdot\mathbf{b} = \sqrt{2}\cdot\sqrt{6}\cos\alpha$$

and also $\mathbf{a}\cdot\mathbf{b} = 3$.

Thus $\cos\alpha = \dfrac{\sqrt{3}}{2}$.

Thus $\alpha = 30°$, and the angle between the plane and the line is 60°.

Linear Dependence

DEFINITION 3.4. *A set of vectors* $\mathbf{v}_1, \mathbf{v}_2, \ldots \mathbf{v}_n$ *is linearly dependent if there are real numbers* $a_1, a_2, \ldots a_n$, *not all zero, such that*

$$a_1\mathbf{v}_1+a_2\mathbf{v}_2+\ldots+a_n\mathbf{v}_n = \mathbf{0},$$

EXAMPLE (xviii). Test for linear dependence the sets:

$$\text{(a)}\ \mathbf{v}_1 = \begin{pmatrix} 2 \\ 1 \\ -1 \end{pmatrix},\ \mathbf{v}_2 = \begin{pmatrix} -6 \\ -3 \\ 3 \end{pmatrix}; \qquad \text{(b)}\ \mathbf{v}_1 = \begin{pmatrix} 0 \\ 0 \end{pmatrix},\ \mathbf{v}_2 = \begin{pmatrix} 7 \\ 5 \end{pmatrix};$$

$$\text{(c)}\ \mathbf{v}_1 = \begin{pmatrix} 3 \\ 3 \end{pmatrix},\ \mathbf{v}_2 = \begin{pmatrix} -2 \\ 0 \end{pmatrix}.$$

(a) Since $3\mathbf{v}_1+\mathbf{v}_2 = \mathbf{0}$, these two vectors are linearly dependent.

(b) Since $\mathbf{v}_1+0\mathbf{v}_2 = \mathbf{0}$, these two vectors are linearly dependent.

(c) If $\qquad a_1\mathbf{v}_1+a_2\mathbf{v}_2 = \mathbf{0},$

then $\qquad 3a_1-2a_2 = 0,$

and $\qquad 3a_1+0a_2 = 0,$

implying that $\qquad a_1 = a_2 = 0.$

Thus these two vectors are not linearly dependent, but *linearly independent*.

Example (xviii) illustrates that two position vectors are linearly dependent if they have parallel directions or if one of them is zero.

EXAMPLE (xix). Test for linear dependence the sets:

$$\text{(a)}\ \mathbf{v}_1 = \begin{pmatrix} 2 \\ 1 \\ -1 \end{pmatrix},\ \mathbf{v}_2 = \begin{pmatrix} -6 \\ -3 \\ 3 \end{pmatrix},\ \mathbf{v}_3 = \begin{pmatrix} 1 \\ 0 \\ 1 \end{pmatrix};$$

$$\text{(b)}\ \mathbf{v}_1 = \begin{pmatrix} 3 \\ 3 \end{pmatrix},\ \mathbf{v}_2 = \begin{pmatrix} -2 \\ 0 \end{pmatrix},\ \mathbf{v}_3 = \begin{pmatrix} 1 \\ -1 \end{pmatrix}; \qquad \text{(c)}\ \mathbf{i},\ \mathbf{j},\ \mathbf{k}.$$

(a) Since $3\mathbf{v}_1+\mathbf{v}_2+0\mathbf{v}_3 = \mathbf{0}$, these vectors are linearly dependent.

(b) If $\qquad a_1\mathbf{v}_1+a_2\mathbf{v}_2+a_3\mathbf{v}_3 = \mathbf{0},$

then $\qquad 3a_1-2a_2+a_3 = 0,$

and $\qquad 3a_1+0a_2-a_3 = 0.$

If $a_1 = 1$, $a_2 = a_3 = 3$, these equations are satisfied.

Thus $\mathbf{v}_1+3\mathbf{v}_2+3\mathbf{v}_3 = \mathbf{0}$, and the vectors are linearly dependent.

(c) If $$a_1\mathbf{i}+a_2\mathbf{j}+a_3\mathbf{k} = \mathbf{0},$$
then $$a_1 = a_2 = a_3 = 0,$$
so that **i**, **j** and **k** are linearly independent.

Note. Example (xix) illustrates that three position vectors are linearly dependent if they are coplanar. (Check that the position vectors in part (a) all lie completely in the plane $x-y+z = 0$.)

Summary of Chapter 3

We have introduced three-dimensional coordinate vectors which obey similar rules to those for two-dimensional vectors for addition, multiplication by a scalar, and for inner product.

We have defined the vector product of two vectors, and deduced a rule for the vector product of coordinate vectors in terms of their coordinates.

We have established the vector equation of a line and its three-dimensional coordinate equations; we have also established the vector equation of a plane and its three-dimensional coordinate equation.

We have defined linear dependence.

Exercise 3b

1. A and B are the points (0, 1, 5), (1, 3, 4). Find the equations of OA and of OB. Find the equations of the line through A parallel to OB, and of the line through B parallel to OA. Find C, the point where these two lines meet. Find the angle between OC and AB. Investigate the shape of $OABC$, and use your discovery to check your result for the angle.

2. The equations of one line are $\dfrac{x-1}{2} = \dfrac{y+2}{4} = \dfrac{z+4}{6}$, and those of another are $x-2 = \dfrac{y}{2} = \dfrac{z+1}{3}$. Show that the two lines are in fact one.

3. Tackle question 6 of Exercise 3a by finding the equations of the line through A and B and testing whether P lies on the line.

4. Find the equation of the plane through the point P (0, 3, 4) perpendicular to OP.

5. Find a vector perpendicular to the plane $x+y+2z = 2$, and one perpendicular to the plane $2x+2z = 5$. Find the perpendicular distance of each plane from O.

6. Find the equations of the line of intersection of the planes of question 5. Find also the angle between the planes.

7. Tackle question 7 of Exercise 3a by finding the equation of the plane through A, B and O and testing whether C lies on thc plane.

8. Show that the line $\frac{x-1}{-2} = \frac{y}{3} = \frac{z-1}{3}$ lies completely in the plane $3x+y+z = 4$.

9. Tackle question 7 of Exercise 3a by testing whether the position vectors $\overline{OA}$, $\overline{OB}$, $\overline{OC}$ are linearly dependent.

10. Tackle Example (ix) by testing whether the vectors $\overline{QP}$, $\overline{QR}$, $\overline{QT}$ are linearly dependent.

CHAPTER 4

GEOMETRY MAPPINGS

Introduction

In this chapter we shall investigate a concept which is widely and fruitfully used in mathematics, that of a mapping. A mapping is a link from one set to another, so that each member of the first set is linked to some unique member of the second set, that unique member being called its image.

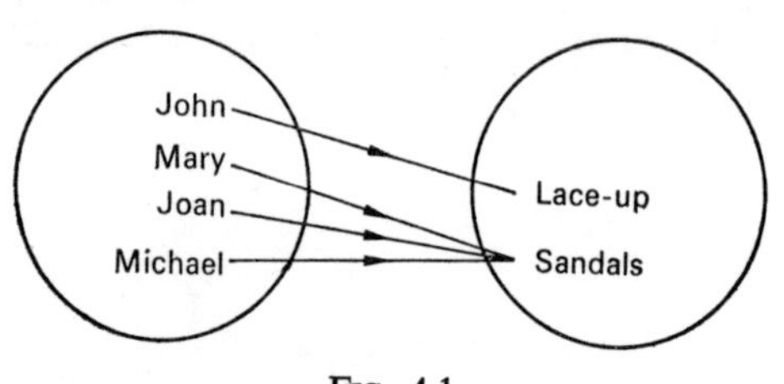

FIG. 4.1

A primary school child might construct a wall chart on the lines of Fig. 4.1. This chart links a set of children's names with a set of shoe types. The links are illustrated by arrows, and denote that on some occasion John was wearing lace-up shoes, Mary sandals, and so on. The link is a mapping from the set of children onto the set of shoe types.

If we reversed the directions of all the arrows in Fig. 4.1, we should not have an illustration of a mapping from the set of shoe types onto the set of children because the shoe type "sandals" would have no image.

Notation. We shall use Greek letters to symbolise mappings. For instance, if ϕ denotes the mapping illustrated in Fig. 4.1,

$$\text{then } \phi(\text{John}) = \text{lace-up shoes.}$$

Or, if ψ denotes the mapping linking every real number to its square, then for any real number x,

$$\psi(x) = x^2.$$

You probably notice that this notation is precisely that of functions. A function is in fact a mapping.

It is sometimes useful to use an arrow notation for mappings:

$$\text{John} \xrightarrow{\phi} \text{lace-up shoes,}$$

$$6 \xrightarrow{\psi} 36.$$

We say that under the mapping ψ, 6 maps onto 36.

EXAMPLE (i). Which of these links are mappings?
(a) The link from every real number to its double,
(b) the link from every real number to its square roots.

(a) Since every real number has a unique image under this link, the link is a mapping. Denoting it by θ,

$$\theta(x) = 2x, \quad \text{for all real } x. \qquad \text{(See Fig. 4.2.)}$$

(b) Since every positive real number has two images, and every negative real number has no image, this link is not a mapping.

Note. The set of elements which are mapped is called the *domain* of the mapping, and the set of images is called the *image set.* The domain and image sets of the mappings we have met are as follows:

Mapping	*Domain*	*Image Set*
ϕ	Set of children	Set of shoe types
ψ	Real numbers	Real numbers
θ	Real numbers	Real numbers

One–one and Many–one Mappings

If each member of the image set of a mapping is the image of only one member of the domain, then that mapping is called *one–one.* But if some member of the image set is the image of more than one member of the domain, that mapping is called *many–one.*

Check that θ is a one–one mapping and that ϕ and ψ are many–one mappings.

Inverse of a Mapping

Suppose α defines some one–one mapping between a domain S and an image set S_1, so that if x is a member of S and x_1 its image in S_1,

$$x_1 = \alpha(x).$$

The mapping which links all such x_1 back to such x is called the inverse of α. It is denoted by α^{-1}, so that

$$x = \alpha^{-1}(x_1).$$

Figure 4.2 illustrates the mapping θ of Example (i) (a) and its inverse mapping θ^{-1}. Notice that many–one mappings do not have inverses; as we observed earlier, if we reversed the direction of the arrows on Fig. 4.1 we should not have the illustration of a mapping.

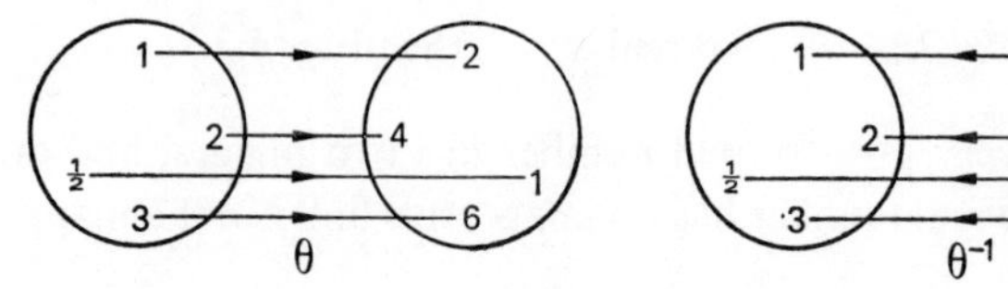

FIG. 4.2

Geometry Mappings

A geometry mapping is a mapping from one set of points to another set of points. For example,

(a) A geographical map is the image set of a domain of points on the surface of the earth. (But a Mercator's Projection, providing no unique image for the North Pole, is not strictly a mapping.)

(b) A reflection in a plane mirror maps the set of points of a domain in front of the mirror onto a set of image points behind it.

(c) A magnification is a mapping.

(d) The rotation of a shape maps the set of points that defined its original position onto a set of points defining its new position. Figure 4.3 shows a domain square *ABCD* and its image $A_1B_1C_1D_1$ after rotation through 45° about the centre of the square.

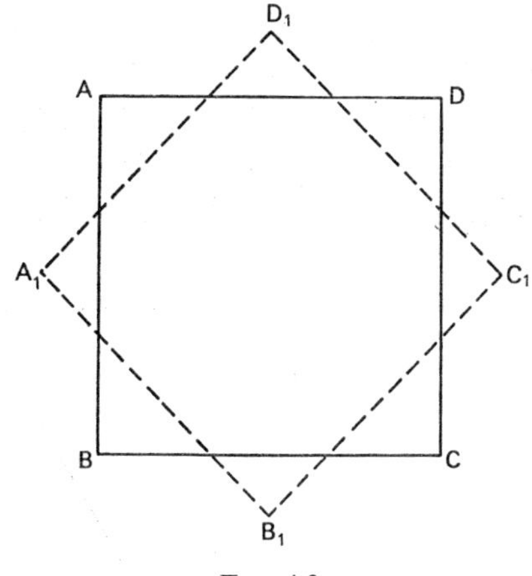

FIG. 4.3

(e) The translation of a shape—that is, moving it without rotating it—is a mapping.

(f) Stretching a sheet of rubber in some random way maps a domain of points on it before the stretch onto an image set (see Fig. 4.4).

(g) Suppose we have a domain of points marked on a sheet of paper. Tearing up the paper maps this domain onto an image set.

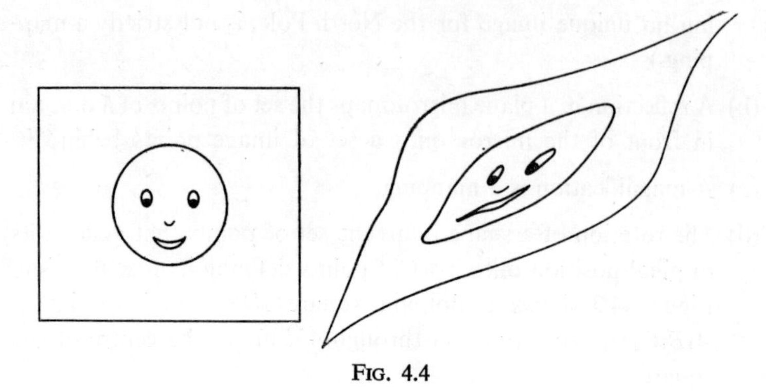

FIG. 4.4

(h) Suppose we have a domain of points, and each point is moved in a certain horizontal direction through a distance proportional to its height above some horizontal plane. Such a movement is a *shear*. Figure 4.5 shows a domain square $ABCD$ and its image $A_1B_1C_1D_1$ after a shear in a direction parallel to BC.

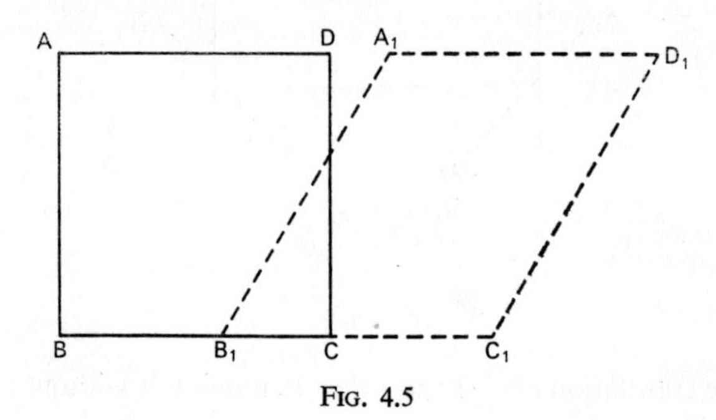

FIG. 4.5

(i) Suppose we have a domain S of points, and each point is linked to the foot of the perpendicular dropped from it onto some plane. Such a link is called a *projection* (see Fig. 4.6).

(j) The link between every point of a domain and itself is a mapping, in which every point is its own image. This link is called the *identity mapping*, and is denoted by ι.

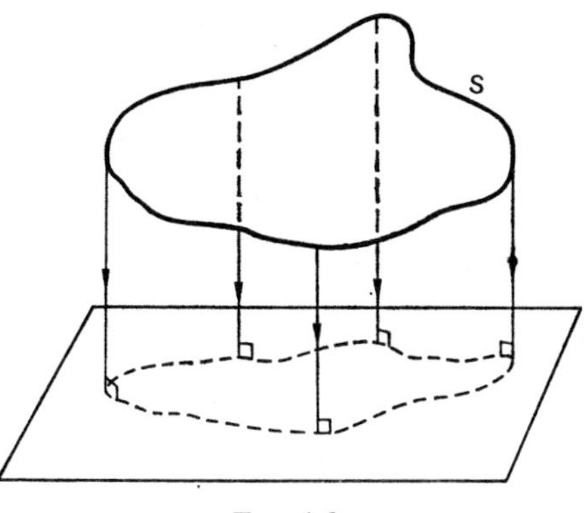

FIG. 4.6

Note. Except for one case, all the mappings (a) to (j) are one–one. Can you describe in words the inverse of each of these one–one mappings? (For instance, the inverse of a reflection is another reflection.)

Invariants

It is interesting to note the properties of a domain that are preserved as properties of the image set. Such properties are called invariants of the mapping. For instance, parallelism of lines is an invariant property of a mapping if lines that are parallel in the domain map onto lines that are parallel in the image set. Connectedness is an invariant property if all points that are connected in the domain map onto connected points in the image set. (Clearly, in our examples, (g) provides the only instance when this property is not preserved.)

Mappings that preserve the parallelism of lines are called *affine mappings.*

Mappings that preserve the connectedness of points are called *topological mappings.*

Mappings that preserve the distances between points are called *isometric mappings.*

Check that among our ten examples of geometry mappings there are seven affine and four isometric mappings. Can you prove that an isometric mapping must also be an affine mapping?

EXAMPLE (ii). Illustrate these mappings of the domain of points forming the x–y-plane:

(a) a reflection in Oy, called the mapping α;

(b) a rotation through 90° about O, called the mapping β;

(c) a projection onto the line Ox, called the mapping γ.

Clearly we cannot illustrate what happens to every point in the x–y-plane under each mapping. Instead, we select a set S of points in the plane, forming some distinctive non-symmetrical shape. The images of S, $\alpha(S)$, $\beta(S)$, $\gamma(S)$ will illustrate what happens to all points under the three mappings.

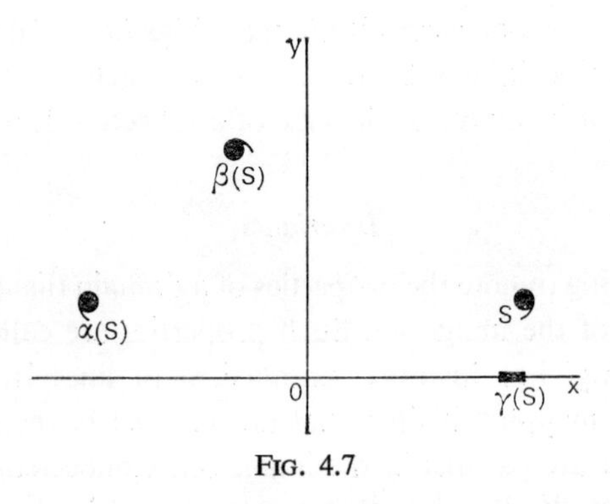

FIG. 4.7

Note on α. We do not pretend that Oy is like a mirror in reflecting only points *in front* of it. Mathematically, any point may be reflected in Oy. The position vector $\begin{pmatrix} x \\ y \end{pmatrix}$ will map onto the position vector $\begin{pmatrix} x_1 \\ y_1 \end{pmatrix}$, where

$$x_1 = -x,$$
$$y_1 = y.$$

Note on β. In most two-dimensional mathematics, we measure a positive angle by an anti-clockwise rotation.

EXAMPLE (iii). Find the inverses of the mappings α, β, γ of Example (ii).

Clearly, $\alpha^{-1} = \alpha$.

β^{-1} is a rotation through $-90°$, or through $270°$. (Whichever of the two rotations is performed, the result is the same, so they are both considered as the same mapping, β^{-1}.)

γ is not a one–one mapping, so it has no inverse.

Note. The inverses of α^{-1} and β^{-1} are α and β.

Product of Mappings

We now give yet another function to that overworked word, *product*. Suppose θ denotes the mapping of a domain S onto an image set S_1, so that for any member x of S there is a unique $x_1 = \theta(x)$ in S_1. Suppose then that φ denotes the mapping from the domain S_1 onto an image set S_2, so that for any x_1 in S_1 there is a unique $x_2 = \varphi(x_1)$ in S_2. Then there is a mapping which maps x in S onto x_2 in S_2, and this mapping is called the product $\varphi\theta$.

Figure 4.8 illustrates the three mappings θ, φ, $\varphi\theta$. Notice that although the mapping order is θ followed by φ, the product is written $\varphi\theta$. This is chosen so that

$$\varphi[\theta(x)] = \varphi\theta(x).$$

(If you are familiar with functional notation, you will be used to this ordering from right to left, in denoting function of a function; for

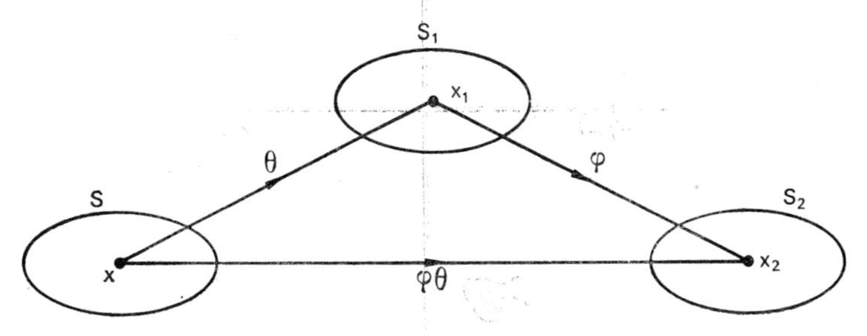

FIG. 4.8

instance, to evaluate log cos x, we first find a number which is cos x, then we find the log of that number.)

Note. In our new algebra of mappings we use notation like that of the algebra of numbers; for instance, the product of the mapping θ with itself is denoted by θ^2. The product of a mapping with its inverse is the identity mapping. Check that for α and β of Examples (ii) and (iii), $\alpha^{-1}\alpha = \beta^{-1}\beta = \iota$.

EXAMPLE (iv). Find the mappings which are the products $\psi\theta$, $\theta\psi$, where θ and ψ are as defined in and before Example (i).

We had $\psi(x) = x^2$ and $\theta(x) = 2x$.

Thus $\psi\theta(x) = \psi(2x) = (2x)^2 = 4x^2$,

and $\theta\varphi(x) = \theta(x^2) = 2x^2$.

Example (iv) illustrates that *the multiplication of mappings is not necessarily a commutative operation.*

EXAMPLE (v). Illustrate the product mappings $\alpha\beta$, $\beta\alpha$, $\alpha\gamma$, $\gamma\alpha$, α^2, β^2, γ^2, where α, β, γ are as defined in Example (ii).

Some of the products are illustrated in Fig. 4.9. Check that $\gamma\alpha = \alpha\gamma$, $\alpha^2 = \gamma^2 = \beta^4 = \iota$.

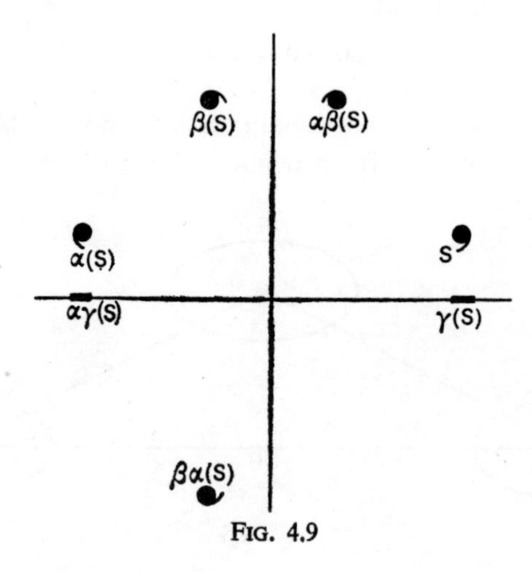

FIG. 4.9

Product of Several Mappings

Suppose θ is a mapping from the set S onto the set S_1, φ from S_1 onto S_2, and ψ from S_2 onto S_3. Then the product mapping $\psi\varphi\theta$ is that mapping which maps any member x of S onto $\psi\{\varphi[\theta(x)]\}$, a

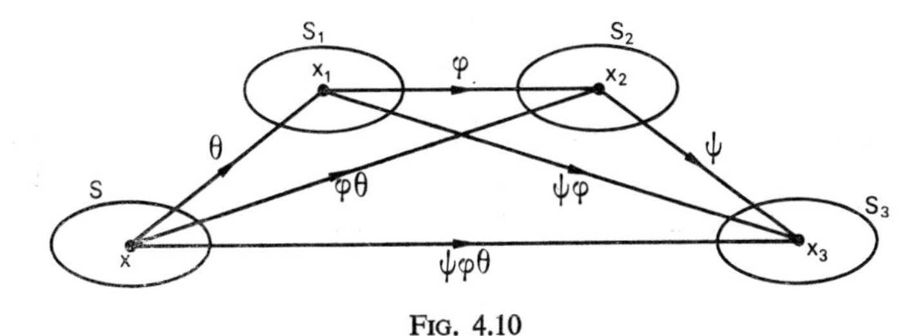

FIG. 4.10

member of S_3. We illustrate these mappings in Fig. 4.10. Also illustrated are the products $\varphi\theta$ and $\psi\varphi$. It is clear from the figure that

$$(\psi\varphi)\theta = \psi(\varphi\theta) = \psi\varphi\theta,$$

that is, *the multiplication of mappings has an associative property.*

Exercise 4a

1. Which of the following links are one–one, many–one, affine, topological or isometric mappings? Discuss the importance of the invariant properties of each mapping:

   (a) the link between the London underground railway system and a map of it,
   (b) the link between two cups of the same tea set,
   (c) the link between a body and its shadow,
   (d) the link between a car and a model of it.

2. Illustrate the following mappings of the x–y-plane:

   (a) α, reflection in the line $y = -x$,
   (b) β, projection onto Ox,
   (c) γ, rotation through 180° about O.
   Find, if they exist, the inverses of α, β and γ.

3. Illustrate mappings which are the products $\alpha\beta$, $\beta\alpha$, $\alpha\gamma$, $\gamma\alpha$, α^2, β^2, γ^2, where α, β, γ are as defined in question 2.

4. Describe in words the mappings which are the products of:
(a) two reflections in different parallel lines,
(b) two reflections in different intersecting lines.

5. Prepare an introduction to mappings in general, or to geometry mappings, for children, bringing in as many aspects as you can of everyday experiences, inventing appropriate games, etc.

Linear Mappings

Some geometry mappings may be defined by linear relations between the coordinates of position vectors. For example, we defined the mapping α of Example (ii) by the relations

$$x_1 = -x,$$
$$y_1 = y$$

between the coordinates of every position vector $\begin{pmatrix} x \\ y \end{pmatrix}$ and its image position vector $\begin{pmatrix} x_1 \\ y_1 \end{pmatrix}$.

In general, if linear relations exist between the coordinates of every position vector and those of its image under some mapping—that is, if

$$x_1 = ax+by,$$
$$y_1 = cx+dy,$$

where a, b, c, d are real numbers—then that mapping is said to be a *linear mapping*.

EXAMPLE (vi). Are these mappings linear: (a) the mapping β, defined in Example (ii), (b) δ, the mapping of the position vector $\begin{pmatrix} x \\ y \end{pmatrix}$ onto the position vector $\begin{pmatrix} x^2 \\ y \end{pmatrix}$?

(a) Under β, the image of $\begin{pmatrix} x \\ y \end{pmatrix}$ is $\begin{pmatrix} x_1 \\ y_1 \end{pmatrix}$, where $x_1 = -y$, $y_1 = x$. That is, β may be defined by linear relations and is thus a linear mapping.

(b) Under δ, the image of $\begin{pmatrix} x \\ y \end{pmatrix}$ is $\begin{pmatrix} x_1 \\ y_1 \end{pmatrix}$, where $x_1 = x^2$, $y_1 = y$.
Thus δ is not a linear mapping.

EXAMPLE (vii). What geometry mappings of the x–y-plane correspond to the following linear relations:

$$\text{(a) } \begin{array}{l} x_1 = 2x, \\ y_1 = 2y; \end{array} \quad \text{(b) } \begin{array}{l} x_1 = x+y, \\ y_1 = y; \end{array} \quad \text{(c) } \begin{array}{l} x_1 = x+y, \\ y_1 = x+y? \end{array}$$

(a) This mapping doubles the length of every vector, leaving its direction invariant.
(b) This mapping shears every vector a distance y in the Ox direction.
(c) Under this mapping, every image position vector has coordinates $\begin{pmatrix} x_1 \\ y_1 \end{pmatrix}$, satisfying the equation $x_1 = y_1$. Thus every point of the x–y-plane maps onto some point on the line whose equation is $y_1 = x_1$. We say that the mapping is onto the line $y_1 = x_1$.

Matrices

Suppose a linear mapping of the x–y-plane is defined by the linear relations:

$$\begin{aligned} x_1 &= ax+by, \\ y_1 &= cx+dy. \end{aligned}$$

The array of coefficients $\begin{pmatrix} a & b \\ c & d \end{pmatrix}$ is called the *matrix* of the mapping.
For example, the matrices of the mappings α, β, γ of Example (ii) are

$$\begin{pmatrix} -1 & 0 \\ 0 & 1 \end{pmatrix}, \quad \begin{pmatrix} 0 & -1 \\ 1 & 0 \end{pmatrix}, \quad \begin{pmatrix} 0 & 0 \\ 0 & 1 \end{pmatrix}.$$

The matrix of the identity mapping of the x–y-plane is

$$\begin{pmatrix} 1 & 0 \\ 0 & 1 \end{pmatrix}.$$

This matrix is called the *identity matrix*.

EXAMPLE (viii). Find the images of the unit vectors **i** and **j** under the mapping of the x–y-plane whose matrix is $\begin{pmatrix} a & b \\ c & d \end{pmatrix}$.

The linear relations defining this mapping are

$$x_1 = ax+by,$$
$$y_1 = cx+dy.$$

To find the image of **i** under the mapping, we must substitute $x = 1$, $y = 0$ in these relations, giving $x_1 = a$, $y_1 = c$. Thus the image of **i** is the position vector $\begin{pmatrix} a \\ c \end{pmatrix}$. Similarly, check that the image of **j** is the position vector $\begin{pmatrix} b \\ d \end{pmatrix}$.

Note. This result is useful in determining the matrix of a geometry mapping. Before we use it in our next example, check that under the mappings α, β, γ of Example (ii), the images of **i** and **j** are in fact those obtained by substituting appropriate values for a, b, c and d in Example (viii).

EXAMPLE (ix). Find the matrices of the following mappings of the x–y-plane:

(a) rotation through an angle θ about O,

(b) reflection in the line $y = x \tan \theta$.

(a) Under this mapping, all position vectors are to be rotated through θ about O. The images $\mathbf{i}_1$, $\mathbf{j}_1$ of the unit vectors **i** and **j** are illustrated in Fig. 4.11. From the figure it is clear that the x coordinate of $\mathbf{i}_1$ is $\cos \theta$ and the y coordinate is $\sin \theta$. Check that this result still applies if θ is an obtuse or a reflex angle.

Thus we establish that under this mapping, $\mathbf{i} \to \begin{pmatrix} \cos \theta \\ \sin \theta \end{pmatrix}$, and we may similarly establish that $\mathbf{j} \to \begin{pmatrix} -\sin \theta \\ \cos \theta \end{pmatrix}$.

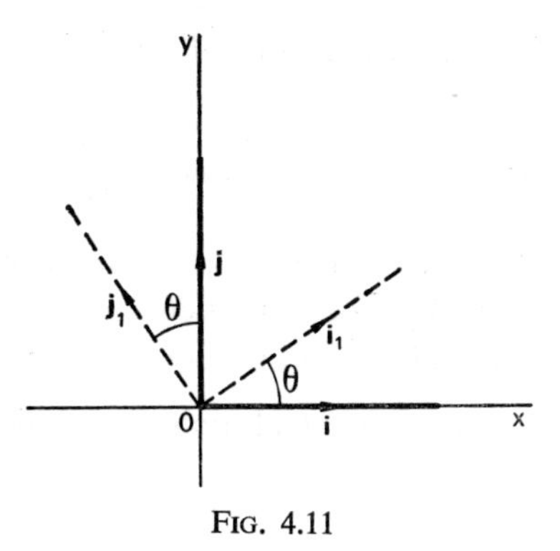

FIG. 4.11

The result of Example (viii) now establishes that the matrix of this mapping is $\begin{pmatrix} \cos\theta & -\sin\theta \\ \sin\theta & \cos\theta \end{pmatrix}$.

(b) Figure 4.12 illustrates the images $\mathbf{i}_2$, $\mathbf{j}_2$ of $\mathbf{i}$ and $\mathbf{j}$ under this mapping. From the figure it is clear that

$$\mathbf{i}_2 = \begin{pmatrix} \cos 2\theta \\ \sin 2\theta \end{pmatrix}, \quad \mathbf{j}_2 = \begin{pmatrix} \cos x \\ -\sin x \end{pmatrix},$$

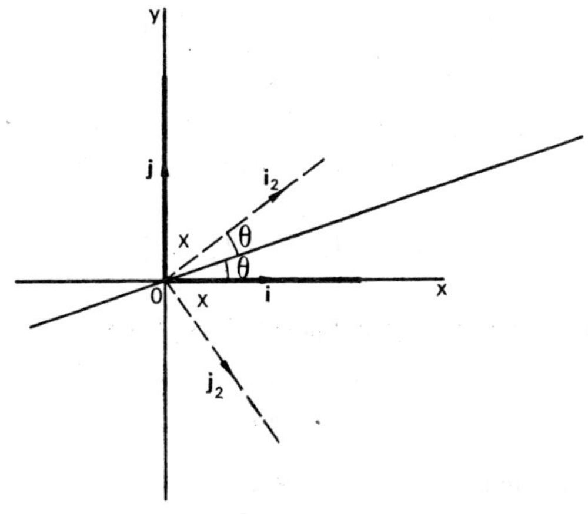

FIG. 4.12

where $x = 90° - 2\theta$. Check that these results still apply if θ is obtuse or reflex.

Thus the matrix of this mapping is $\begin{pmatrix} \cos 2\theta & \sin 2\theta \\ \sin 2\theta & -\cos 2\theta \end{pmatrix}$.

We shall now investigate what happens to some particular two-dimensional curves under some particular linear mappings. We assume that you are familiar with the coordinate equations of:

(a) a circle, centre O, radius a, all points on which satisfy

$$x^2+y^2 = a^2,$$

(b) an ellipse, centre O, major axis along Ox, $2a$ in length, minor axis along Oy, $2b$ in length, all points on which satisfy

$$\frac{x^2}{a^2}+\frac{y^2}{b^2} = 1,$$

(c) a hyperbola, centre O, major axis along Ox, $2a$ in length, minor axis along Oy, $2b$ in length, all points on which satisfy

$$\frac{x^2}{a^2}-\frac{y^2}{b^2} = 1.$$

(A brief consultation with any sixth-form textbook on coordinate geometry will familiarise you with the definitions of these curves and the derivation of their equations.)

EXAMPLE (x). In Fig. 4.13 find the images of the circle C, whose equation is $x^2+y^2 = 1$, under the mapping whose matrices are

$$\text{(a) } \begin{pmatrix} 2 & 0 \\ 0 & 2 \end{pmatrix}, \quad \text{(b) } \begin{pmatrix} 2 & 0 \\ 0 & 1 \end{pmatrix}.$$

(a) The linear relations defining this mapping are

$$x_1 = 2x,$$
$$y_1 = 2y,$$

implying that

$$x = \tfrac{1}{2}x_1,$$
$$y = \tfrac{1}{2}y_1.$$

Thus, if x and y satisfy the equation $x^2+y^2 = 1$, then x_1 and y_1 satisfy $\frac{x_1^2}{4}+\frac{y_1^2}{4} = 1$. That is, the image of C is C_1, a concentric circle of radius 2.

(b) This mapping is defined by the relations

$$x_2 = 2x,$$
$$y_2 = y.$$

Thus if $x^2+y^2 = 1$, then $\frac{x_2^2}{4}+y_2^2 = 1$. That is, the image of C is C_2, an ellipse of major axis 4, minor axis 2.

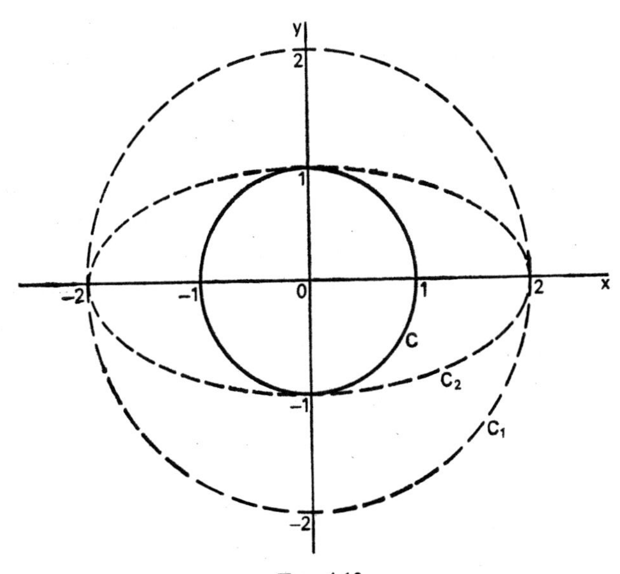

FIG. 4.13

EXAMPLE (xi). Find the image of the straight line L, whose equation is $y = mx+k$, under the mapping whose matrix is $\begin{pmatrix} a & b \\ c & d \end{pmatrix}$.

The mapping is defined by the relations

$$x_1 = ax+by,$$
$$y_1 = cx+dy,$$

implying that

$$(ad-bc)x = dx_1-by_1,$$
$$(ad-bc)y = ay_1-cx_1.$$

Thus if $y = mx+k$, then

$$ay_1-cx_1 = m(dx_1-by_1)+k(ad-bc),$$

or

$$(a+b)y_1 = (md+c)x_1+k(ad-bc).$$

That is, the image of L is L_1, another straight line.

EXAMPLE (xii). Find the images of the unit square U, whose vertices are the points (0, 0), (1, 0), (0, 1), (1, 1), under the mappings whose matrices are

$$\text{(a)} \begin{pmatrix} 2 & 0 \\ 0 & 1 \end{pmatrix}, \quad \text{(b)} \begin{pmatrix} 1 & 1 \\ 1 & 1 \end{pmatrix}, \quad \text{(c)} \begin{pmatrix} 2 & 1 \\ 1 & 1 \end{pmatrix}.$$

(a) Under this mapping, the images of the four points are clearly respectively (0, 0), (2, 0), (0, 1), (2, 1).

In Example (xi) we have shown that all straight lines map onto straight lines. Thus the image U_1 of U is a rectangle, as in Fig. 4.14.

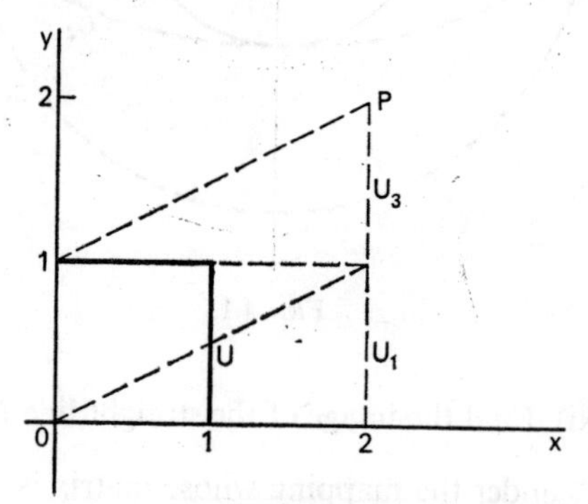

FIG. 4.14

(b) The images of the four points are clearly (0, 0), (1, 1), (1, 1), (2, 2). Thus the image, U_2, of U is the line segment OP, where P is the point (2, 2).

(c) The images of the four points are (0, 0), (2, 1), (1, 1), (3, 2). Thus the image, U_3, of U is a parallelogram, as in Fig. 4.14.

Three-dimensional Mappings

We now consider geometry mappings of the domain of points forming x–y–z space. Such a mapping is linear if linear relations exist between the coordinates of every position vector

$$\begin{pmatrix} x \\ y \\ z \end{pmatrix} \quad \text{and its image} \quad \begin{pmatrix} x_1 \\ y_1 \\ z_1 \end{pmatrix}$$

of the form

$$\begin{aligned} x_1 &= ax+by+cz, \\ y_1 &= dx+ey+fz, \\ z_1 &= gx+hy+qz, \end{aligned}$$

where the coefficients a, b, c, etc., are real numbers. The array of coefficients

$$\begin{pmatrix} a & b & c \\ d & e & f \\ g & h & q \end{pmatrix}$$

is called the *matrix* of the mapping. It is easily shown, analogously to the two-dimensional case, that under the mapping defined by this matrix, the images of **i**, **j** and **k** are the position vectors

$$\begin{pmatrix} a \\ d \\ g \end{pmatrix}, \quad \begin{pmatrix} b \\ e \\ h \end{pmatrix}, \quad \begin{pmatrix} c \\ f \\ q \end{pmatrix}.$$

In three dimensions, reflection takes place in a plane, and as we pointed out in the two-dimensional case, mathematically we do not distinguish between the *front* and *back* of the plane, but consider all points to be reflected in it. A rotation takes place about a line, called the *axis of rotation*.

EXAMPLE (xiii). Find the matrices of the following mappings of x–y–z space, given that they are linear mappings:

(a) rotation through 180° about Ox,
(b) reflection in the plane $x = 0$.

(a) Under this mapping, $\mathbf{i} \to \mathbf{i}$, $\mathbf{j} \to -\mathbf{j}$, $\mathbf{k} \to \mathbf{k}$.

Thus the matrix of the mapping is $\begin{pmatrix} 1 & 0 & 0 \\ 0 & -1 & 0 \\ 0 & 0 & -1 \end{pmatrix}$.

(b) Under this mapping, $\mathbf{i} \to -\mathbf{i}$, $\mathbf{j} \to \mathbf{j}$, $\mathbf{k} \to \mathbf{k}$.

Thus the matrix of the mapping is $\begin{pmatrix} -1 & 0 & 0 \\ 0 & 1 & 0 \\ 0 & 0 & 1 \end{pmatrix}$.

The Central Quadrics

We end the chapter with a note on these surfaces, whose cross-sections are always conic sections, and whose centres are at the origin.

The Sphere. Any point on the surface of the sphere centre O and radius a has coordinates (x, y, z) which satisfy

$$x^2+y^2+z^2 = a^2.$$

Conversely, any point whose coordinates satisfy this equation lie on the surface of the sphere. The equation is called the *equation of the sphere*.

The Ellipsoid. Consider the surface consisting of all points whose coordinates satisfy

$$\frac{x^2}{a^2}+\frac{y^2}{b^2}+\frac{z^2}{c^2} = 1.$$

The points on the surface whose z coordinates are k must satisfy

$$\frac{x^2}{a^2}+\frac{y^2}{b^2} = 1-\frac{k^2}{c^2},$$

that is, they must lie on an ellipse. In other words, provided $k < c$, the plane $z = k$ cuts this surface in an ellipse.

Similarly we can show that the planes $x = l$, $y = m$ cut the surface in ellipses. The surface is called an *ellipsoid*, and it is illustrated in Fig. 4.15.

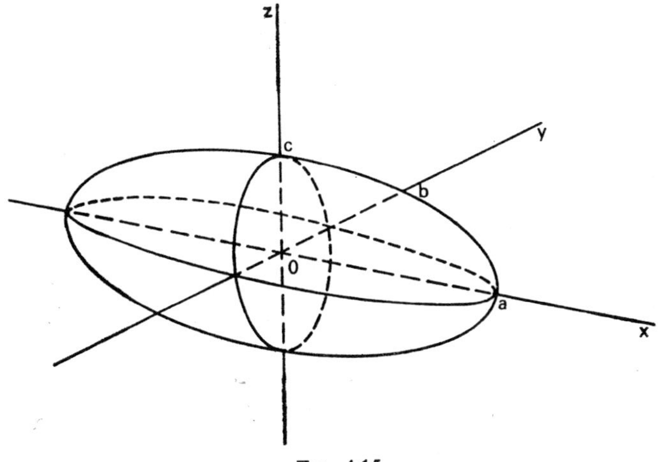

FIG. 4.15

The Hyperboloid of One Sheet. Consider the surface consisting of all points whose coordinates satisfy

$$\frac{x^2}{a^2}+\frac{y^2}{b^2}-\frac{z^2}{c^2} = 1.$$

The points on the surface whose z coordinates are k must satisfy

$$\frac{x}{a^2}+\frac{y}{b^2} = 1+\frac{k^2}{c^2},$$

that is, the plane $z = k$ cuts this surface in an ellipse.

But the planes $x = l$, $y = m$ cut the surface in hyperbolas. The surface is called a *hyperboloid of one sheet*, and it is illustrated in Fig. 4.16.

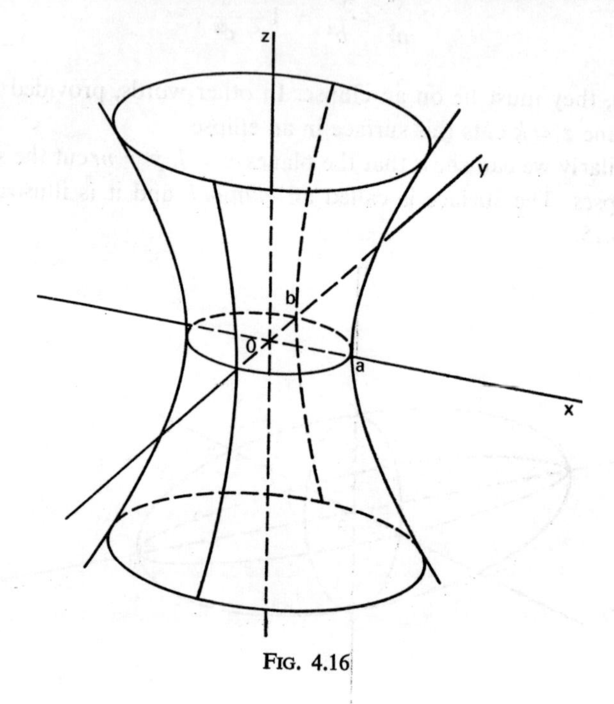

FIG. 4.16

The Hyperboloid of Two Sheets. Consider the surface consisting of all points whose coordinates satisfy

$$-\frac{x^2}{a^2}-\frac{y^2}{b^2}+\frac{z^2}{c^2}=1.$$

The points on the surface whose z coordinates are zero must satisfy

$$-\frac{x^2}{a^2}-\frac{y^2}{b^2}=1.$$

Clearly, since the quantity x^2 must be positive or zero, no points have coordinates satisfying this equation, implying that no point on this surface has a z coordinate of zero.

However, the plane $z = k$ cuts the surface in points whose co-ordinates satisfy

$$\frac{x^2}{a^2}+\frac{y^2}{b^2} = \frac{k^2}{c^2}-1.$$

If $k^2 > c^2$, this is the equation of an ellipse.

The planes $x = l$, $y = m$ cut the surface in hyperbolas. The surface is called a *hyperboloid of two sheets*, being in two parts, and it is illustrated in Fig. 4.17.

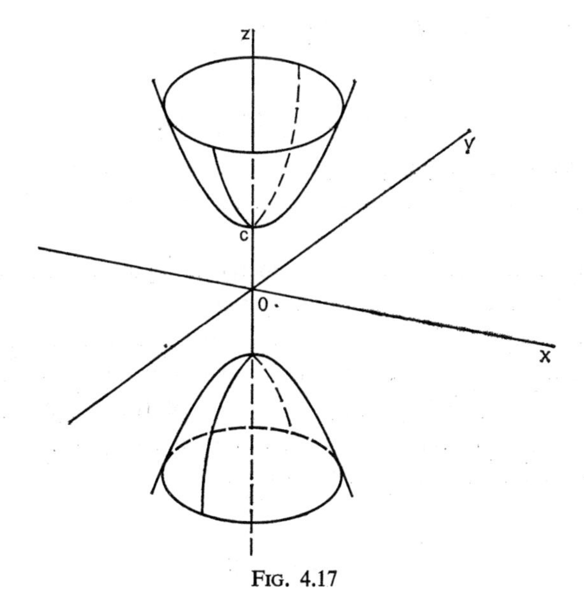

FIG. 4.17

EXAMPLE (xiv). Find the image of the sphere S, whose equation is

$$x^2+y^2+z^2 = 1,$$

under the mapping whose matrix is

$$\begin{pmatrix} 3 & 0 & 0 \\ 0 & 2 & 0 \\ 0 & 0 & 1 \end{pmatrix}.$$

This mapping is defined by the relations

$$x_1 = 3x,$$
$$y_1 = 2y,$$
$$z_1 = z.$$

Thus, if $x^2+y^2+z^2 = 1$, then

$$\frac{x_1^2}{9}+\frac{y_1^2}{4}+z_1^2 = 1.$$

Thus the image of S is an ellipsoid.

Summary of Chapter 4

We have defined a mapping between a domain and an image set. A mapping may be one–one, which has an inverse, or many–one, which has no inverse.

We have met examples of geometry mappings, including some which are affine, topological and isometric.

We have defined the product of mappings and seen that this operation has an associative but not necessarily a commutative property.

We have investigated linear mappings of the x–y-plane and of x–y–z-space, and defined the matrix of such a mapping.

We have introduced the central quadrics.

Exercise 4b

1. Write down the vectors which are the images of the position vectors **i**, **j** and $\mathbf{v} = \begin{pmatrix} x \\ y \end{pmatrix}$ under the following mappings of the x–y-plane:

(a) reflection in Ox,
(b) reflection in the line $y = x$,
(c) rotation through 180° about O,
(d) rotation through 90° about O,
(e) a shear of y in the direction of Ox,
(f) projection onto Oy.

Write down the matrix of each mapping. Use the result of Example (viii) if this helps.

2. Find the images of the unit square and the circle $x^2+y^2 = 25$ under the mapping whose matrix is $\begin{pmatrix} 3 & 0 \\ 0 & 2 \end{pmatrix}$.

Guess at a matrix that maps the ellipse $\dfrac{x^2}{4}+\dfrac{y^2}{9} = 1$ onto a circle.

3. Describe in words the mappings of the x–y-plane whose matrices are:

$$\text{(a)} \begin{pmatrix} 3 & 0 \\ 0 & 3 \end{pmatrix}, \quad \text{(b)} \begin{pmatrix} 1 & 0 \\ 2 & 1 \end{pmatrix}, \quad \text{(c)} \begin{pmatrix} 2 & 2 \\ 1 & 1 \end{pmatrix},$$

and the mappings of x–y–z-space whose matrices are:

$$\text{(d)} \begin{pmatrix} 1 & 0 & 0 \\ 0 & 1 & 0 \\ 0 & 0 & -1 \end{pmatrix}, \quad \text{(e)} \begin{pmatrix} -1 & 0 & 0 \\ 0 & 1 & 0 \\ 0 & 0 & -1 \end{pmatrix}, \quad \text{( )} \begin{pmatrix} 0 & 1 & 0 \\ 1 & 0 & 0 \\ 0 & 0 & 1 \end{pmatrix}.$$

4. Find the images of the unit cube and the sphere $x^2+y^2+z^2 = 9$ under the mapping whose matrix is

$$\begin{pmatrix} 1 & 0 & 0 \\ 0 & 2 & 0 \\ 0 & 0 & 3 \end{pmatrix}.$$

Guess at a matrix that maps the ellipsoid $\dfrac{x^2}{4}+\dfrac{y^2}{9}+z^2 = 1$ onto a sphere.

5. Show that in two dimensions the reflection in the line $y = x$ of the curve $y = f(x)$ has the equation $x_1 = f(y_1)$. Illustrate by sketching graphs of $y = e^x$, $y_1 = \log x_1$, $y = \sin x$, $y_1 = \arcsin x_1$, $y = x^2$, $y_1 = \pm\sqrt{x_1}$.

CHAPTER 5

CLASSIFICATION OF TWO BY TWO MATRICES

Introduction

We have seen that not every mapping of the x–y-plane is linear. In this chapter we shall investigate some properties of those mappings that are linear, and classify them according to the types of geometry mappings which they define.

Consider first a general linear mapping defined by the matrix $M = \begin{pmatrix} a & b \\ c & d \end{pmatrix}$. In Example (xi) of Chapter 4 we saw that this matrix maps the straight line L, whose equation is $y = mx+k$, onto another straight line, L_1. Notice that the gradient of the line L_1 was dependent on the values of m, a, b, c and d, but not on k.

Now consider any line parallel to L. Its equation must be of the form $y = mx+l$, where l is some real number. The matrix M will map this line onto an image line, whose gradient is dependent on m, a, b, c and d, but not on l. In fact, this image line will be parallel to L_1.

The significant fact to emerge is that under any linear mapping of the x–y-plane, parallelism is preserved. In other words, all such mappings are *affine*.

Area and Determinants

Under the mapping defined by the matrix M, parallelism is preserved. In particular, the unit square will map into a parallelogram, as illustrated by $OABC$ in Fig. 5.1.

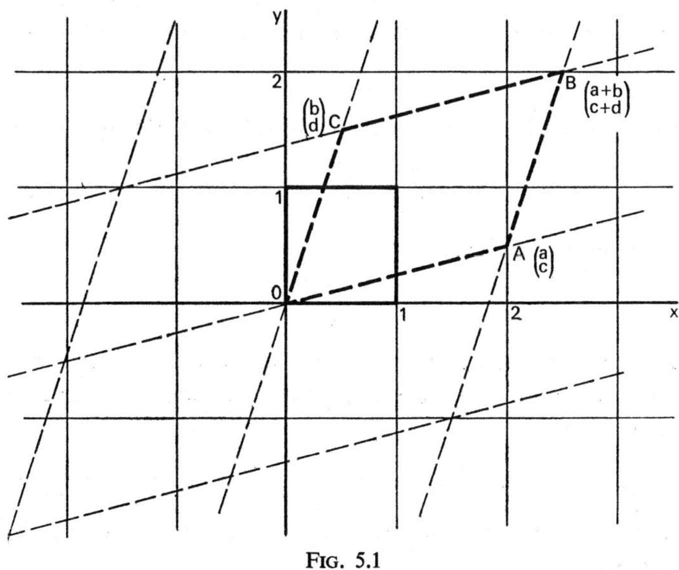

FIG. 5.1

EXAMPLE (i). Find the area of the image parallelogram $OABC$ of Fig. 5.1.

Referring to Example (xi) of Chapter 2, in which we calculated the area of a triangle, we deduce that the area $OABC$ is $|ad-bc|$.

The quantity $ad-bc$ will have a significant place in much of our work with the matrix $\begin{pmatrix} a & b \\ c & d \end{pmatrix}$, and it has a name.

DEFINITION 5.1. *Given* $M = \begin{pmatrix} a & b \\ c & d \end{pmatrix}$, *the quantity* $ad-bc$ *is called the determinant of the matrix* M. *It may be denoted by* det M.

Note. A network of unit squares covering the x–y-plane will map under M onto a network of parallelograms all congruent to $OABC$. The area of a plane surface is usually measured by the number of unit squares it encloses, but it could equally well be measured using a parallelogram as a unit of area. Since under M a shape enclosing n unit squares will map onto a shape enclosing n image parallelograms, we deduce that the mapping multiplies the area of any shape by $|\det M|$.

EXAMPLE (ii). Find the area enclosed by the ellipse whose equation is

$$\frac{x^2}{4}+\frac{y^2}{9} = 1.$$

The linear mapping

$$x_1 = \tfrac{1}{2}x,$$
$$y_1 = \tfrac{1}{3}y,$$

maps this ellipse onto the circle whose equation is $x_1^2+y_1^2 = 1$. (You may rightly have conjectured this in answering question 2 of Exercise 4b.)

The matrix of this mapping is $\begin{pmatrix} \frac{1}{2} & 0 \\ 0 & \frac{1}{3} \end{pmatrix}$, whose determinant is $\frac{1}{6}$. Therefore, under the mapping, the area of the ellipse has been multiplied by $\frac{1}{6}$.

But the area of the circle is π, so that the area of the ellipse must be 6π.

Isometric Mappings and Orthogonal Matrices

Some of the mappings we examined in Chapter 4 were both affine and isometric. For a mapping of the x–y-plane to be isometric, the image of the unit square must itself be a unit square. If the matrix $M = \begin{pmatrix} a & b \\ c & d \end{pmatrix}$ defines the mapping, certain conditions must be satisfied by a, b, c, and d for the mapping to be isometric. The images of the vectors **i** and **j** under the mapping are $\begin{pmatrix} a \\ c \end{pmatrix}$ and $\begin{pmatrix} b \\ d \end{pmatrix}$ respectively, and these image vectors must be perpendicular unit vectors. A matrix with this property has a name.

DEFINITION 5.2. *A two by two matrix whose columns are perpendicular unit vectors is an orthogonal matrix.*

You may have met the word *orthogonal* before in connection with geometry, when it means *perpendicular*.

Examples of orthogonal matrices include

$$\begin{pmatrix} 0 & 1 \\ 1 & 0 \end{pmatrix}, \quad \begin{pmatrix} \dfrac{1}{\sqrt{2}} & \dfrac{-1}{\sqrt{2}} \\ \dfrac{1}{\sqrt{2}} & \dfrac{1}{\sqrt{2}} \end{pmatrix}, \quad \begin{pmatrix} \cos\theta & -\sin\theta \\ \sin\theta & \cos\theta \end{pmatrix},$$

and all have the form $\begin{pmatrix} a & b \\ c & d \end{pmatrix}$, where $a^2+c^2 = b^2+d^2 = 1$ (unit vectors), and $ab+cd = 0$ (inner product of perpendicular vectors).

EXAMPLE (iii). Find the image of the curve C, whose equation is

$$5x^2-4xy+2y^2 = 1,$$

under the mapping defined by the matrix

$$U = \begin{pmatrix} \dfrac{1}{\sqrt{5}} & \dfrac{2}{\sqrt{5}} \\ \dfrac{2}{\sqrt{5}} & \dfrac{-1}{\sqrt{5}} \end{pmatrix}.$$

Deduce the shape of C.

As this matrix is orthogonal, it defines an isometry, so that the image curve C_1 will have the same shape as C.

The mapping is defined by the relations

$$\sqrt{5}x_1 = x+2y,$$
$$\sqrt{5}y_1 = 2x-y.$$

Eliminating y from the two relations gives $x = \dfrac{1}{\sqrt{5}}(x_1+2y_1)$, and further algebra gives $y = \dfrac{1}{\sqrt{5}}(2x_1-y_1)$.

Thus we deduce from the relationship $5x^2-4xy+2y^2 = 1$ that

$$\tfrac{5}{5}(x_1+2y_1)^2-\tfrac{4}{5}(x_1+2y_1)(2x_1-y_1)+\tfrac{2}{5}(2x_1-y_1)^2 = 1.$$

That is,

$$x_1^2+6y_1^2 = 1,$$

showing that C_1, and hence C, must be an ellipse, whose semi-axes are 1 and $\dfrac{1}{\sqrt{6}}$.

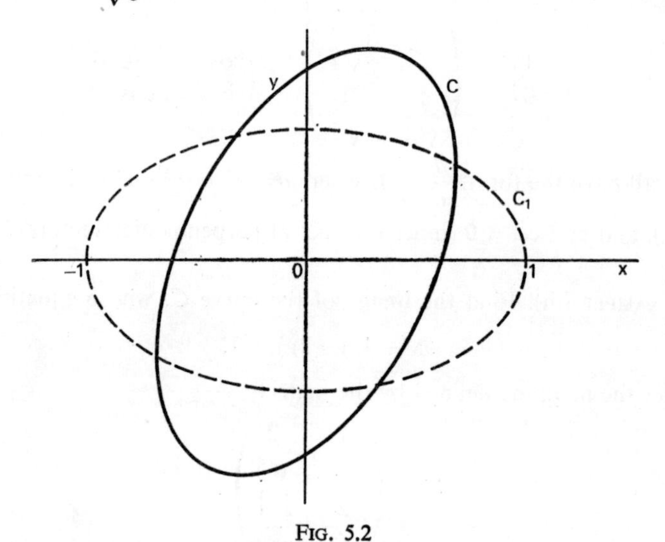

FIG. 5.2

Many–One Mappings and Singular Matrices

We have seen that under the mapping of the x–y-plane defined by the matrix M, the area enclosed by any shape is multiplied by $|\det M|$. What is implied if $\det M = 0$? Can a shape enclosing an area map onto a shape enclosing no area? We saw in Example (vii), part (c), of Chapter 4 that this is indeed possible; there the unit square mapped onto a line segment under the matrix $\begin{pmatrix}1 & 1\\ 1 & 1\end{pmatrix}$. Notice that the determinant of this matrix is zero.

DEFINITION 5.3. *The matrix M is singular if* $\det M = 0$.

EXAMPLE (iv). Describe the geometry mappings whose matrices are:

$$\text{(a)} \begin{pmatrix} 1 & 2 \\ 2 & 4 \end{pmatrix}, \quad \text{(b)} \begin{pmatrix} 1 & 4 \\ 0 & 0 \end{pmatrix}, \quad \text{(c)} \begin{pmatrix} 0 & 0 \\ 0 & 0 \end{pmatrix}.$$

(a) The mapping is defined by the relations

$$x_1 = x+2y,$$
$$y_1 = 2x+4y.$$

Thus for all x and y, $y_1 = 2x_1$, implying that every image point lies on the straight line $y_1 = 2x_1$.

(b) Every image point lies on the straight line $y_1 = 0$.

(c) Every image point is at the origin O.

Note. Example (iv) illustrates that a singular matrix defines a many–one mapping onto a straight line through the origin, or onto the origin itself.

A further deduction about singular matrices may be drawn. We could have tackled Example (iv) by considering the columns of each matrix, since these form vectors which are the edges of the image parallelogram of the unit square. Since in each case these vectors are linearly dependent (see Definition 3.4), we deduce that they are collinear and therefore that the image parallelogram is a line segment.

Example (iv) in fact *illustrates that a singular two by two matrix has rows which are linearly dependent and columns which are linearly dependent.*

Inverse Mappings and Inverse Matrices

In Chapter 4 we introduced the concept of inverse mappings, and saw that only one–one mappings have inverses. Now we shall define the inverse of a matrix.

DEFINITION 5.4. *If the matrix M defines the one–one mapping of the x–y plane, θ, then the matrix defining the mapping θ^{-1} is called the inverse matrix of M, and is denoted by M^{-1}.*

EXAMPLE (v). Find, if they exist, the inverses of the following matrices:

$$\text{(a) } M_1 = \begin{pmatrix} 1 & 1 \\ 0 & 1 \end{pmatrix}, \quad \text{(b) } M_2 = \begin{pmatrix} 2 & 0 \\ 0 & 2 \end{pmatrix}, \quad \text{(c) } M_3 = \begin{pmatrix} 1 & 1 \\ 1 & 1 \end{pmatrix}.$$

(a) M_1 defines a shear of amount y in the Ox direction. The inverse of this mapping is a shear of amount $-y$ in the Ox direction. (See Example (vii) of Chapter 4.)

The matrix of this inverse mapping is $M_1^{-1} = \begin{pmatrix} 1 & -1 \\ 0 & 1 \end{pmatrix}$.

(b) M_2 defines a magnification with a factor of 2. The inverse mapping is a reduction to half size, or magnification with a factor of $\frac{1}{2}$.

The matrix of this inverse mapping is $M_2^{-1} = \begin{pmatrix} \frac{1}{2} & 0 \\ 0 & \frac{1}{2} \end{pmatrix}$.

(c) M_3 defines a many-one mapping, which has no inverse. Thus M_3 has no inverse matrix.

We sometimes find a matrix in which every entry has a common factor, such as the matrix U of Example (iii) of this chapter. The following definition provides a neater notation for such a matrix, which will also lead to simpler computations in future.

DEFINITION 5.5. *Every entry of the matrix kM is k times the corresponding entry of the matrix M.*

For example, if $M = \begin{pmatrix} a & b \\ c & d \end{pmatrix}$, then $kM = \begin{pmatrix} ka & kb \\ kc & kd \end{pmatrix}$; and the matrix U of Example (iii) may be written $\frac{1}{\sqrt{5}}\begin{pmatrix} 1 & 2 \\ 2 & -1 \end{pmatrix}$. Check that the matrix M_2 of Example (v) is $2I$, and that its inverse M_2^{-1} is $\frac{1}{2}I$.

Exercise 5a

1. Show that $\det I = 1$. What is $\det kI$? Find a relation between $\det kM$ and $\det M$.

2. By finding a linear mapping which maps it onto a circle, find the area enclosed by the ellipse whose equation is

$$\frac{x^2}{a^2}+\frac{y^2}{b^2} = 1.$$

Explain why no linear mapping of the x–y-plane will map the hyperbola

$$\frac{x^2}{a^2}-\frac{y^2}{b^2} = 1$$

onto a circle.

3. Show that the matrix $U = \frac{1}{5}\begin{pmatrix}3 & 4\\ 4 & -3\end{pmatrix}$ is orthogonal. This matrix maps the curve C, whose equation is

$$7x^2-48xy-7y^2 = 1$$

onto the curve C_1. Find the equation of C_1 and deduce the shape of C.

4. Classify these matrices and describe the geometry mappings of the x–y-plane defined by each:

$$\text{(a) } M_1 = \begin{pmatrix}-1 & 2\\ 2 & -4\end{pmatrix}, \quad \text{(b) } M_2 = \begin{pmatrix}\cos 60° & -\sin 60°\\ \sin 60° & \cos 60°\end{pmatrix},$$

$$\text{(c) } M_3 = \begin{pmatrix}\cos 60° & \sin 60°\\ \sin 60° & -\cos 60°\end{pmatrix}.$$

Reference to Example (ix) of Chapter 4 may help for parts (b) and (c).

5. Find the inverses, if they exist, of the matrices of question 4. Verify in each case that

$$\det(M^{-1}) = \frac{1}{\det M}.$$

6. Show that the determinant of any two by two orthogonal matrix is ± 1. (*Hint:* consider the mapping defined by such a matrix and its effect on area.)

Product of Mappings and Product of Matrices

In Chapter 4 we defined the product of two mappings. On this we shall build a definition of the product of two matrices.

EXAMPLE (vi). The matrix $M_1 = \begin{pmatrix} 0 & -1 \\ 1 & 0 \end{pmatrix}$ defines the mapping α, and the matrix $M_2 = \begin{pmatrix} -1 & 0 \\ 0 & 1 \end{pmatrix}$ defines the mapping β. Show that the mapping $\beta\alpha$ is also a linear mapping, and find its matrix.

The mapping α is defined by the relations

$$x_1 = -y,\quad y_1 = x.$$

β, following α, maps all $\begin{pmatrix} x_1 \\ y_1 \end{pmatrix}$ onto $\begin{pmatrix} x_2 \\ y_2 \end{pmatrix}$, where

$$x_2 = -x_1,\quad y_2 = y_1.$$

Thus the mapping $\beta\alpha$ maps all $\begin{pmatrix} x \\ y \end{pmatrix}$ onto $\begin{pmatrix} x_2 \\ y_2 \end{pmatrix}$, where

$$x_2 = y,\quad y_2 = x.$$

Thus $\beta\alpha$ is a linear mapping whose matrix is $\begin{pmatrix} 0 & 1 \\ 1 & 0 \end{pmatrix}$.

Example (vi) illustrates that the product of two linear mappings is a linear mapping. We now show that this is generally true.

Suppose that the mapping θ, whose matrix is $M = \begin{pmatrix} a & b \\ c & d \end{pmatrix}$, is followed by the mapping ϕ, whose matrix is $N = \begin{pmatrix} \alpha & \beta \\ \gamma & \delta \end{pmatrix}$. That is,

θ maps all $\begin{pmatrix} x \\ y \end{pmatrix}$ onto $\begin{pmatrix} x_1 \\ y_1 \end{pmatrix}$, where

$$x_1 = ax+by,\quad y_1 = cx+dy,$$

and ϕ maps all $\begin{pmatrix} x_1 \\ y_1 \end{pmatrix}$ onto $\begin{pmatrix} x_2 \\ y_2 \end{pmatrix}$, where

$$x_2 = \alpha x_1+\beta y_1,\quad y_2 = \gamma x_1+\delta y_1.$$

Then $\phi\theta$ maps all $\begin{pmatrix} x \\ y \end{pmatrix}$ onto $\begin{pmatrix} x_2 \\ y_2 \end{pmatrix}$, and by substituting in the above relations to eliminate x_1 and y_1, we have the relations

$$x_2 = (\alpha a+\beta c)x+(\alpha b+\beta d)y,$$
$$y_2 = (\gamma a+\delta c)x+(\gamma b+\delta d)y.$$

This establishes that the product mapping is linear.

We can now define the product of two matrices.

DEFINITION 5.6. *If M, N are the matrices of the mappings of the x–y-plane θ, ϕ, respectively, then the matrix of the product mapping $\phi\theta$ is called the product of the matrices and is denoted by NM.*

Rule for Computing the Product of Matrices. From the previous working it is clear that for

$$N = \begin{pmatrix} \alpha & \beta \\ \gamma & \delta \end{pmatrix}, \quad M = \begin{pmatrix} a & b \\ c & d \end{pmatrix},$$

$$NM = \begin{pmatrix} \alpha a+\beta c & \alpha b+\beta d \\ \gamma a+\delta c & \gamma b+\delta d \end{pmatrix}.$$

This gives us a rule for computing the product which is quite easily recognised, but not easily put into words. It has been expressed with more vividness than accuracy as: "Shoot along the rows of N and dive down the columns of M, multiplying entries and adding as you go!"

Check that, with reference to Example (vi), $M_2M_1 = \begin{pmatrix} 0 & 1 \\ 1 & 0 \end{pmatrix}$.

Note. As with mapping multiplication, we omit any symbol to represent the operation of matrix multiplication. From our experience with mappings, we see that matrix multiplication is associative, but not necessarily commutative.

EXAMPLE (vii). Compute the products M_1M_2, M_1^2, M_2^2, $M_1M_1^{-1}$, $M_1^{-1}M_1$, where M_1 and M_2 are the matrices of Example (vi). Interpret the results geometrically.

$$M_1M_2 = \begin{pmatrix} -1 & 0 \\ 0 & 1 \end{pmatrix}\begin{pmatrix} 0 & -1 \\ 1 & 0 \end{pmatrix} = \begin{pmatrix} 0 & 1 \\ 1 & 0 \end{pmatrix}.$$

Geometrically, M_1 defines a rotation through 90° about O, and M_2 defines a reflection in the Oy axis. The matrix M_1M_2 defines a reflection in the line $y = x$, which is the result of a reflection in the Oy axis followed by a rotation through 90° about O.

$$M_1^2 = \begin{pmatrix} 0 & -1 \\ 1 & 0 \end{pmatrix}\begin{pmatrix} 0 & -1 \\ 1 & 0 \end{pmatrix} = \begin{pmatrix} -1 & 0 \\ 0 & -1 \end{pmatrix}.$$

Geometrically, M_1^2 defines a rotation through 180° about O, which is the result of a rotation through 90° about O followed by itself.

$$M_2^2 = \begin{pmatrix} -1 & 0 \\ 0 & 1 \end{pmatrix}\begin{pmatrix} -1 & 0 \\ 0 & 1 \end{pmatrix} = \begin{pmatrix} 1 & 0 \\ 0 & 1 \end{pmatrix}.$$

M_2^2 is the identity matrix, defining the identity mapping, the result of a reflection in Oy followed by itself. This establishes that M_2 is its own inverse—a property of the matrix of any reflection.

M_1^{-1} defines a rotation through $-90°$, and is thus $\begin{pmatrix} 0 & 1 \\ -1 & 0 \end{pmatrix}$.

Check that $M_1 M_1^{-1} = M_1^{-1} M_1 = I$.

Example (vii) illustrates that a non-singular matrix commutes with its inverse, the product being the identity matrix.

Transpose of a Matrix and Transpose of a Vector

It is sometimes convenient to have a name and a notation for a particular kind of rearrangement of a matrix.

DEFINITION 5.7. *The transpose of the matrix M is the matrix M', which is formed by writing the columns of M as the corresponding rows of M'.*

Thus, if $M = \begin{pmatrix} a & b \\ c & d \end{pmatrix}$, then $M' = \begin{pmatrix} a & c \\ b & d \end{pmatrix}$.

DEFINITION 5.8. *The matrix A is symmetric if $A' = A$.*

Examples of symmetric matrices include

$$\begin{pmatrix} 1 & 0 \\ 0 & 1 \end{pmatrix}, \quad \begin{pmatrix} 3 & -2 \\ -2 & 6 \end{pmatrix}, \quad \begin{pmatrix} 0 & 0 \\ 0 & 5 \end{pmatrix}.$$

EXAMPLE (viii). Show that for any two by two matrix M, the matrix $M'M$ is symmetric. Deduce that if M is orthogonal, then $M'M = I$.

$$\text{Let } M = \begin{pmatrix} a & b \\ c & d \end{pmatrix}. \text{ Then } M'M = \begin{pmatrix} a & c \\ b & d \end{pmatrix}\begin{pmatrix} a & b \\ c & d \end{pmatrix}$$

$$= \begin{pmatrix} a^2+c^2 & ab+cd \\ ab+cd & b^2+d^2 \end{pmatrix}.$$

Thus $M'M$ is symmetric.

If M is orthogonal, then $a^2+c^2 = b^2+d^2 = 1$, and $ab+cd = 0$. (See Definition 5.2 *et seq.*)

In such a case, $M'M = I$.

Note. Example (viii) establishes that if M is an orthogonal matrix, then M' is the inverse of M.

DEFINITION 5.9. *The transpose of the vector* $\mathbf{v} = \begin{pmatrix} x \\ y \end{pmatrix}$ *is denoted by* $\mathbf{v}' = (x \quad y)$. $\mathbf{v}'$ *is called a row vector.*

At the moment we attach no geometric significance to the vector $\mathbf{v}'$, but we shall apply to such row vectors similar rules for addition and multiplication by a scalar to those we employ for column vectors. There is no meaning to be attached, however, to the sum of a row vector and a column vector.

We also extend the rule for computing the product of two matrices to apply in a similar way to the product of a matrix and a vector. For example, if $M = \begin{pmatrix} a & b \\ c & d \end{pmatrix}$ and $\mathbf{v} = \begin{pmatrix} x \\ y \end{pmatrix}$, we compute the product $M\mathbf{v}$ thus: the first row of M combines with the column of $\mathbf{v}$ to produce an entry $ax+by$; then the second row of M combines with the column of $\mathbf{v}$ to produce $cx+dy$. $M\mathbf{v}$ will thus have only one column, and will be a column vector. We may write

$$\begin{pmatrix} a & b \\ c & d \end{pmatrix}\begin{pmatrix} x \\ y \end{pmatrix} = \begin{pmatrix} ax+by \\ cx+dy \end{pmatrix}.$$

In fact, if $\mathbf{v}_1 = \begin{pmatrix} x_1 \\ y_1 \end{pmatrix}$, then the linear relations

$$x_1 = ax+by,$$
$$y_1 = cx+dy$$

may be expressed concisely by the vector equation

$$M\mathbf{v} = \mathbf{v}_1.$$

Notice that we cannot, however, apply our product rule to compute $\mathbf{v}M$, for the "rows" of $\mathbf{v}$ do not contain the same number of entries as the columns of M. The product $\mathbf{v}M$ is said to be undefined.

Other products which may be computed are:

$$\mathbf{v}'M = (x \quad y)\begin{pmatrix} a & b \\ c & d \end{pmatrix} = (ax+by \quad cx+dy),$$

the product here being a row vector, and

$$\mathbf{v}'\mathbf{v}_1 = (x \quad y)\begin{pmatrix} x_1 \\ y_1 \end{pmatrix} = xx_1+yy_1,$$

the product here being a "one by one matrix"—that is, a scalar quantity. You will remember this scalar quantity as the inner product $\mathbf{v} \cdot \mathbf{v}_1$, which we defined in Definition 2.2.

Check that $I\mathbf{v} = \mathbf{v}$, and that $\mathbf{v}'\mathbf{v} = v^2$.

Having extended the operation of matrix multiplication to apply to these new situations, have we lost the associative property of matrix multiplication? Let us test whether $N(M\mathbf{v})$ is the same as $(NM)\mathbf{v}$. $N(M\mathbf{v})$ is the vector which is the image of the vector $M\mathbf{v}$ under the mapping whose matrix is N. This means that it is the image of $\mathbf{v}$ under the mapping M followed by the mapping of N. By the definition of product of matrices, this is $(NM)\mathbf{v}$. Thus multiplication of matrix and vector is associative.

Example (ix). Given $\mathbf{v} = \begin{pmatrix} x \\ y \end{pmatrix}$, $\mathbf{v}_1 = \begin{pmatrix} x_1 \\ y_1 \end{pmatrix}$, $M = \begin{pmatrix} a & b \\ c & d \end{pmatrix}$, and $M\mathbf{v} = \mathbf{v}_1$, show that (a) $\mathbf{v}_1' = \mathbf{v}'M'$, and (b) $\mathbf{v} = M^{-1}\mathbf{v}_1$, provided M^{-1} exists.

(a) $M\mathbf{v} = \mathbf{v}_1$ implies that

$$x_1 = ax+by,$$
$$y_1 = cx+dy.$$

Thus $\quad \mathbf{v}'M' = (x \quad y)\begin{pmatrix} a & c \\ b & d \end{pmatrix} = (x_1 \quad y_1) = \mathbf{v}_1'.$

(b) $M\mathbf{v} = \mathbf{v}_1$.

Multiply both sides of this equality on the left by M^{-1}. Then

$$M^{-1}(M\mathbf{v}) = M^{-1}\mathbf{v}_1.$$

By the associative property, $M^{-1}(M\mathbf{v}) = (M^{-1}M)\mathbf{v}$, and since also $(M^{-1}M)\mathbf{v} = I\mathbf{v} = \mathbf{v}$, the equality may be rewritten

$$\mathbf{v} = M^{-1}\mathbf{v}_1.$$

We conclude by solving Example (iii) again by a new method, replacing some of the traditional algebra used in our former solution with some matrix algebra. The problem was to find the image of the curve C, whose equation is

$$5x^2-4xy+2y^2 = 1,$$

under the mapping defined by the orthogonal matrix $U = \dfrac{1}{\sqrt{5}}\begin{pmatrix} 1 & 2 \\ 2 & -1 \end{pmatrix}$.

The mapping is defined by the vector relation

$$U\mathbf{v} = \mathbf{v}_1,$$

where $\quad \mathbf{v} = \begin{pmatrix} x \\ y \end{pmatrix}$ and $\mathbf{v}_1 = \begin{pmatrix} x_1 \\ y_1 \end{pmatrix}$.

By the result of Example (ix) (b),

$$\mathbf{v} = U^{-1}\mathbf{v}_1.$$

Since U is orthogonal, $U^{-1} = U'$ (Example (vii)), and since U is symmetric, $U' = U$.

Thus $\quad \mathbf{v} = U\mathbf{v}_1.$

This implies the relations $x = \dfrac{1}{\sqrt{5}}x_1+\dfrac{2}{\sqrt{5}}y_1,$

$$y = \frac{2}{\sqrt{5}}x_1-\frac{1}{\sqrt{5}}y_1.$$

The method continues from this point as in Example (iii).

Summary of Chapter 5

We discovered that all linear mappings of the x–y-plane are affine, and that under the mapping whose matrix is M, the area of any shape in the plane is multiplied by $|\det M|$.

We defined an orthogonal matrix and saw that such a matrix defines an isometric mapping.

We defined a singular matrix, and saw that such a matrix defines a many–one mapping.

We defined the inverse of a matrix via the inverse of a mapping.

We defined the product of matrices, and saw that matrix multiplication has an associative but not necessarily a commutative property.

We defined the transpose of a matrix and the transpose of a vector, and extended the rule for the product of matrices to apply to the product of matrix and vector, and to the product of vector and vector.

We showed that with this extension of the rule, the associative property of matrix multiplication is preserved.

We proved that if U is an orthogonal matrix, then $U^{-1} = U'$.

Exercise 5b

1. Find the matrix of the mapping which is equivalent to a reflection of the x–y-plane in the Ox axis, followed by a reflection in the line $y = x$,
(a) geometrically, using the result of Exercise 4a, question 4,
(b) by multiplying the appropriate matrices.

2. X is the matrix $\begin{pmatrix} 1 & 1 \\ -1 & -1 \end{pmatrix}$. What geometry mappings of the x–y-plane do X and X^2 define? Explain the apparent paradox of your answers.

3. The matrices M_2, M_3 were defined in question 4 of Exercise 5a. Compute the products M_2M_3, M_3M_2, M_3^2, and interpret your results geometrically, checking that each product matrix defines a mapping which is the product of two other appropriate mappings.

4. In Example (ix) of Chapter 4, we found the matrix defining a rotation of the x–y-plane through an angle θ about O. Write down the matrix which defines a rotation through an angle $(\theta+\phi)$ about O. Demonstrate that this matrix must be the product of two other matrices of the same type. Compute this product, and deduce a formula giving $\cos(\theta+\phi)$ in terms of $\cos\theta$, $\sin\theta$, $\cos\phi$, $\sin\phi$. Deduce a similar formula for $\sin(\theta+\phi)$.

5. In question 5 of Exercise 5a, you found the inverses of the matrices M_2, M_3. Check that $M_2^{-1} = M_2'$. What type of matrices are M_2, M_3? Verify by calculation that the products $M_2M_2^{-1}$, $M_2^{-1}M_2$ are both equal to I.

6. Conjecture, if you can, a formula for the inverse M^{-1} of the matrix $M = \begin{pmatrix} a & b \\ c & d \end{pmatrix}$. Your conjecture will be correct if it confirms that $M^{-1}M = I$, and it should show that a singular matrix has no inverse.

7. D is the matrix $\begin{pmatrix} a & 0 \\ 0 & b \end{pmatrix}$. Find D^2, D^3, and deduce a formula for D^n, where n is a positive integer. Does your formula remain valid if n has the value -1?

8. M is the matrix $\begin{pmatrix} 1 & 2 \\ 3 & 4 \end{pmatrix}$, $\mathbf{v}$ the vector $\begin{pmatrix} x \\ y \end{pmatrix}$. Form, where possible, the products $M\mathbf{v}$, $M'\mathbf{v}$, $M\mathbf{v}'$, $M'\mathbf{v}'$, $\mathbf{v}'M$, $\mathbf{v}'M'$, $\mathbf{v}M$, $\mathbf{v}M'$, $M'M$, MM', $\mathbf{v}\mathbf{v}'$, $\mathbf{v}'\mathbf{v}$.

9. Show that multiplication of a matrix by a vector is distributive over vector addition. That is, show that

$$M\mathbf{v}+M\mathbf{w} = M(\mathbf{v}+\mathbf{w}),$$

where $M = \begin{pmatrix} a & b \\ c & d \end{pmatrix}$, $\quad \mathbf{v} = \begin{pmatrix} x \\ y \end{pmatrix}$, $\quad$ and $\quad \mathbf{w} = \begin{pmatrix} X \\ Y \end{pmatrix}$.

10. Establish the following general results for two by two matrices, M, N:

(a) $(MN)^{-1} = N^{-1}M^{-1}$,
(b) $(MN)' = N'M'$.

CHAPTER 6

CLASSIFICATION OF THREE BY THREE MATRICES

Introduction

In this chapter we shall consider three by three matrices and the geometry mappings of x–y–z-space which they define. We shall classify the matrices as we did in Chapter 5.

As in the two-dimensional case, it can be shown that a linear mapping of x–y–z-space maps straight lines onto straight lines. It can also be shown that parallelism is preserved, implying that all such mappings are affine.

In particular, the unit cube will map onto a solid with six faces, each of which is a parallelogram. Such a solid is called a parallelepiped (see Fig. 6.1).

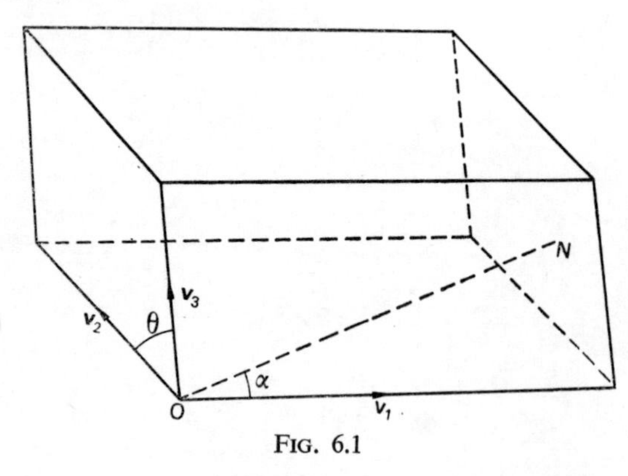

FIG. 6.1

Volume and Determinants

EXAMPLE (i). Find the volume of the image parallelepiped of the unit cube under the mapping whose matrix is

$$M = \begin{pmatrix} a & b & c \\ d & e & f \\ g & h & q \end{pmatrix}.$$

Let $\mathbf{v}_1 = \begin{pmatrix} a \\ d \\ g \end{pmatrix}$, and let $\mathbf{v}_2$, $\mathbf{v}_3$ denote the other columns of M.

Every edge of the image parallelepiped must be parallel to one of $\mathbf{v}_1$, $\mathbf{v}_2$ or $\mathbf{v}_3$.

The volume of a parallelepiped is the product of the area of one face and the perpendicular distance between it and the face parallel to it.

The area of the face edged by $\mathbf{v}_2$ and $\mathbf{v}_3$ is $|v_2 v_3 \sin\theta|$, where θ is the angle between $\mathbf{v}_2$ and $\mathbf{v}_3$. This quantity is the size of the vector product $\mathbf{v}_2 \times \mathbf{v}_3$ (defined in Definition 3.1). If ON is perpendicular to this face (as in Fig. 6.1), and $\mathbf{n}$ is a unit vector in the direction of ON, then

$$\mathbf{v}_2 \times \mathbf{v}_3 = (v_2 v_3 \sin\theta)\mathbf{n}.$$

ON is in fact the perpendicular distance between the two parallel faces edged by $\mathbf{v}_2$ and $\mathbf{v}_3$; and since the length ON may be expressed as $\mathbf{v}_1 \cdot \mathbf{n}$, the volume of the parallelepiped can be expressed as $|\mathbf{v}_1 \cdot (\mathbf{v}_2 \times \mathbf{v}_3)|$.

Now, $$\mathbf{v}_2 \times \mathbf{v}_3 = \begin{pmatrix} eq-fh \\ hc-bq \\ bf-ce \end{pmatrix},$$

so that $$|\mathbf{v}_1 \cdot (\mathbf{v}_2 \times \mathbf{v}_3)| = |a(eq-fh)+d(hc-bq)+g(bf-ce)|.$$

Note. The quantity obtained in Example (i) will have a significant place in much of our work with three by three matrices, and it has a name, the *determinant* of the matrix M. As in the two by two case, it is denoted by det M.

Notice that

$$\det M = a \det \begin{pmatrix} e & f \\ h & q \end{pmatrix} - d \det \begin{pmatrix} b & c \\ h & q \end{pmatrix} + g \det \begin{pmatrix} b & c \\ e & f \end{pmatrix}.$$

This expression is called the expansion of det M by its first column. Each of its three terms is the product of an element of the first column of M and the two by two determinant formed by the entries of M that are not in the same row or column as that element.

Notice also that

$$\det M = a \det \begin{pmatrix} e & f \\ h & q \end{pmatrix} - b \det \begin{pmatrix} d & f \\ g & q \end{pmatrix} + c \det \begin{pmatrix} d & e \\ g & h \end{pmatrix}.$$

This is the expansion of det M by its first row.

EXAMPLE (ii). Evaluate det U, where $U = \frac{1}{3}\begin{pmatrix} 2 & 1 & -2 \\ 1 & 2 & 2 \\ 2 & -2 & 1 \end{pmatrix}$.

Expanding det U by its first column gives

$$\det U = \tfrac{2}{3} \det \begin{pmatrix} \frac{2}{3} & \frac{2}{3} \\ -\frac{2}{3} & \frac{1}{3} \end{pmatrix} - \tfrac{1}{3} \det \begin{pmatrix} \frac{1}{3} & -\frac{2}{3} \\ -\frac{2}{3} & \frac{1}{3} \end{pmatrix} + \tfrac{2}{3} \det \begin{pmatrix} \frac{1}{3} & \frac{2}{3} \\ \frac{2}{3} & \frac{2}{3} \end{pmatrix}$$

$$= \tfrac{2}{3}(\tfrac{2}{3}\cdot\tfrac{1}{3}+\tfrac{2}{3}\cdot\tfrac{2}{3}) - \tfrac{1}{3}(\tfrac{1}{3}\cdot\tfrac{1}{3}+\tfrac{2}{3}\cdot\tfrac{2}{3}) + \tfrac{2}{3}(\tfrac{1}{3}\cdot\tfrac{2}{3}+\tfrac{2}{3}\cdot\tfrac{2}{3})$$

$$= 1.$$

Note. A network of unit cubes covering x–y–z-space will map under M onto a network of congruent parallelepipeds. The volume of a solid is usually measured by the number of unit cubes it encloses, but it could equally well be measured using a parallelepiped as a unit of volume. Since under M a shape enclosing n unit cubes will map onto a shape n image parallelepipeds, we deduce that the mapping multiplies the volume of any shape by $|\det M|$.

EXAMPLE (iii). Find the volume enclosed by the ellipsoid whose equation is

$$\frac{x^2}{4} + \frac{y^2}{9} + z^2 = 1.$$

The linear mapping

$$x_1 = \tfrac{1}{2}x,$$
$$y_1 = \tfrac{1}{3}y,$$
$$z_1 = z$$

maps this ellipsoid onto the sphere whose equation is

$$x^2+y^2+z^2 = 1.$$

(You may rightly have conjectured this in answering question 4 of Exercise 4b.)

The matrix of this mapping is $\begin{pmatrix} \frac{1}{2} & 0 & 0 \\ 0 & \frac{1}{3} & 0 \\ 0 & 0 & 1 \end{pmatrix}$, whose determinant is $\frac{1}{6}$. Therefore the mapping has multiplied the volume of the ellipsoid by $\frac{1}{6}$.

The volume of the sphere is $\frac{4}{3}\pi$, so that the volume of the ellipsoid must be 8π.

Isometric Mappings and Orthogonal Matrices

Definition 5.2, concerning a two by two orthogonal matrix, may be extended to apply to a three by three matrix. A three by three matrix whose columns are mutually orthogonal, or perpendicular, unit vectors is an orthogonal matrix. Examples of such matrices include

$$\begin{pmatrix} 1 & 0 & 0 \\ 0 & 0 & 1 \\ 0 & -1 & 0 \end{pmatrix}, \quad \begin{pmatrix} \frac{1}{\sqrt{2}} & \frac{-1}{\sqrt{2}} & 0 \\ \frac{1}{\sqrt{2}} & \frac{1}{\sqrt{2}} & 0 \\ 0 & 0 & 1 \end{pmatrix}, \quad \frac{1}{3}\begin{pmatrix} 2 & 1 & -2 \\ 1 & 2 & 2 \\ 2 & -2 & 1 \end{pmatrix},$$

and all have the form $\begin{pmatrix} a & b & c \\ d & e & f \\ g & h & q \end{pmatrix}$, where

$$a^2+d^2+g^2 = b^2+e^2+h^2 = c^2+f^2+q^2 = 1,$$

and

$$ab+de+gh = bc+ef+hq = ac+df+gq = 0.$$

Since the mapping defined by an orthogonal matrix U is an isometry, the volume enclosed by any shape will be unaltered by the mapping. Therefore $|\det U| = 1$. Check that the above examples of orthogonal matrices have this property.

EXAMPLE (iv). Find the image of the surface S, whose equation is

$$41x^2+75y^2+34z^2-24xz = 25,$$

under the mapping defined by the matrix $U = \frac{1}{5}\begin{pmatrix} 3 & 0 & 4 \\ 0 & 5 & 0 \\ 4 & 0 & -3 \end{pmatrix}$.

U, being orthogonal, defines an isometry, so that the image surface of S, S_1, say, will have the same shape as S.

The mapping is defined by the relations

$$\begin{aligned} 5x_1 &= 3x+4z, \\ y_1 &= y, \\ 5z_1 &= 4x-3z. \end{aligned}$$

Solving these for x, y and z gives

$$\begin{aligned} x &= 3x_1+4z_1, \\ y &= y_1, \\ z &= 4x_1-3z_1. \end{aligned}$$

Thus we deduce from the relationship

$$41x^2+75y^2+34z^2-24xz = 25$$

that

$$\tfrac{41}{25}(3x_1+4z_1)^2+75y_1^2+\tfrac{34}{25}(4x_1-3z_1)^2-\tfrac{24}{25}(3x_1+4z_1)(4x_1-3z_1) = 25.$$

That is,
$$x_1^2+3y_1^2+2z_1^2 = 1,$$

showing that S_1, and hence S, is an ellipsoid whose semi-axes are 1, $\dfrac{1}{\sqrt{3}}$ and $\dfrac{1}{\sqrt{2}}$.

Many–One Mappings and Singular Matrices

If det M $= 0$ for a three by three matrix M, M is said to be singular. Then, under the mapping defined by M, the image parallelepiped of the unit cube has zero volume, and M defines a many–one mapping which may be onto the origin, onto a straight line, or onto a plane.

EXAMPLE (v). Describe the geometry mappings of x–y–z-space defined by

$$\text{(a) } M_1 = \begin{pmatrix} 1 & 2 & -1 \\ 2 & 4 & -2 \\ 1 & 0 & 3 \end{pmatrix}, \quad \text{(b) } M_2 = \begin{pmatrix} 1 & 2 & -1 \\ 2 & 4 & -2 \\ 3 & 6 & -3 \end{pmatrix}.$$

(a) The mapping is defined by the relations

$$\begin{aligned} x_1 &= x+2y-z, \\ y_1 &= 2x+4y-2z, \\ z_1 &= x+3z. \end{aligned}$$

Thus for all x, y and z, $y_1 = 2x_1$, implying that every image point lies on the plane $y_1 = 2x_1$.

(b) For all x, y and z, $y_1 = 2x_1$ and $z_1 = 3x_1$, implying that every image point lies on the straight line $x_1 = \dfrac{y_1}{2} = \dfrac{z_1}{3}$.

Note. We could have tackled Example (v) by considering the columns of each matrix, since these form vectors which are the edges of the image parallelepiped of the unit cube. It is clear that the columns of M_1 are linearly dependent, and we saw in Chapter 3 that if three position vectors are linearly dependent, then they are coplanar. We deduce that the image parallelepiped must degenerate to a parallelogram. In the case of M_2, any two of the columns are clearly linearly dependent, so we deduce that any two of such position vectors are collinear. That is, they are all collinear, and the image parallelepiped degenerates to a line segment.

Example (v) in fact illustrates that a *singular three by three matrix has rows which are linearly dependent and columns which are linearly dependent.*

Inverse Mappings and Inverse Matrices

Definition 5.4 may be amended to apply to three by three matrices by replacing the term *x–y-plane* by *x–y–z-space.*

EXAMPLE (vi). Find, if they exist, the inverses of the following matrices:

$$\text{(a) } M_1 = \begin{pmatrix} 0 & -1 & 0 \\ 1 & 0 & 0 \\ 0 & 0 & 1 \end{pmatrix}, \quad \text{(b) } M_2 = \begin{pmatrix} 0 & 1 & 0 \\ 1 & 0 & 0 \\ 0 & 0 & 1 \end{pmatrix},$$

$$\text{(c) } M_3 = \begin{pmatrix} 1 & 2 & -1 \\ 2 & 0 & 2 \\ -1 & -3 & 2 \end{pmatrix}.$$

(a) Under M_1, $\mathbf{i}$ maps onto $\mathbf{j}$, $\mathbf{j}$ maps onto $-\mathbf{i}$, $\mathbf{k}$ maps onto $\mathbf{k}$. Therefore M_1 defines a rotation through 90° about Oz. The inverse mapping must be a rotation through 90° in the opposite sense. So under M_1^{-1}, $\mathbf{i} \to -\mathbf{j}$, $\mathbf{j} \to \mathbf{i}$, and $\mathbf{k} \to \mathbf{k}$.

Thus
$$M_1^{-1} = \begin{pmatrix} 0 & 1 & 0 \\ -1 & 0 & 0 \\ 0 & 0 & 1 \end{pmatrix}.$$

(b) Under M_2, $\mathbf{i} \to \mathbf{j}$, $\mathbf{j} \to \mathbf{i}$, and $\mathbf{k} \to \mathbf{k}$. Therefore M_2 defines a reflection in the plane $y = x$. Every reflection is its own inverse, so $M_2^{-1} = M_2$.

(c) The columns (and rows) of M_3 are linearly dependent. Therefore M_3 is a singular matrix and has no inverse.

Note. It is not always simple to investigate the geometry mapping defined by a three by three matrix. In Chapter 8 we shall establish a method of finding the inverse of a matrix which is not dependent on knowing the geometry mapping associated with the matrix. Until then, we shall confine our need for inverse matrices to those whose geometry mappings are easily recognisable.

Exercise 6a

1. I is the identity three by three matrix. Show that $\det I = 1$. If Definition 5.5 may be extended to apply to three by three matrices, what is $\det kI$? Find a relation between $\det kM$ and $\det M$, where M is any three by three matrix.

2. By finding a matrix which maps it onto a sphere, find the volume enclosed by the ellipsoid whose equation is

$$\frac{x^2}{a^2}+\frac{y^2}{b^2}+\frac{z^2}{c^2} = 1.$$

Explain why no linear mapping will map the hyperboloid

$$\frac{x^2}{a^2}+\frac{y^2}{b^2}-\frac{z^2}{c^2} = 1$$

onto a sphere.

3. Show that the matrix $U = \frac{1}{3}\begin{pmatrix} 2 & -1 & 2 \\ -1 & 2 & 2 \\ 2 & 2 & -1 \end{pmatrix}$ is orthogonal. Find the image of the surface, S, whose equation is

$$6x^2+7y^2+5z^2-4xz+4xy = 3,$$

under the mapping defined by U. Deduce the shape of S.

4. Classify these matrices, and describe the geometry mappings of x–y–z-space defined by each:

(a) $M_1 = \begin{pmatrix} -1 & 2 & 4 \\ 0 & 1 & 6 \\ 2 & -4 & -8 \end{pmatrix}$, $\quad M_2 = \frac{1}{2}\begin{pmatrix} \sqrt{3} & 0 & -1 \\ 0 & 2 & 0 \\ 1 & 0 & \sqrt{3} \end{pmatrix}$, $\quad M_3 = \begin{pmatrix} 1 & 2 & 3 \\ 0 & 0 & 0 \\ 1 & 2 & 3 \end{pmatrix}$.

Compute the determinant of each matrix.

5. Find the inverse of the matrix M_2 of question 4. Verify that

$$\det(M_2^{-1}) = \frac{1}{\det M_2}.$$

6. M is the matrix $\begin{pmatrix} 1 & -1 & 1 \\ 2 & 3 & 7 \\ 3 & 2 & 8 \end{pmatrix}$. Show that M is singular, by

(a) showing that its rows are linearly dependent,
(b) showing that its columns are linearly dependent,
(c) evaluating det M.

7. Show that the determinant of any three by three orthogonal matrix is ± 1. (*Hint:* consider mappings.)

Product of Mappings and Product of Matrices

Definition 5.6 and the rule for multiplying matrices which followed it apply to three by three matrices, as do the properties of associativity and commutativity between a matrix and its inverse.

EXAMPLE (vii). Compute the products M_1M_2, M_2M_1, M_1^2, M_2^2, where M_1, M_2 are as defined in Example (vi). Interpret the results geometrically.

$$M_1M_2 = \begin{pmatrix} 0 & -1 & 0 \\ 1 & 0 & 0 \\ 0 & 0 & 1 \end{pmatrix}\begin{pmatrix} 0 & 1 & 0 \\ 1 & 0 & 0 \\ 0 & 0 & 1 \end{pmatrix} = \begin{pmatrix} -1 & 0 & 0 \\ 0 & 1 & 0 \\ 0 & 0 & 1 \end{pmatrix}.$$

Geometrically, M_1M_2 defines a reflection in the plane $x = 0$, which is the result of a reflection in the plane $y = x$ followed by a rotation through 90° about Oz.

$$M_2M_1 = \begin{pmatrix} 1 & 0 & 0 \\ 0 & -1 & 0 \\ 0 & 0 & 1 \end{pmatrix}.$$

M_2M_1 defines a reflection in the plane $y = 0$, which is the result of a rotation through 90° about Oz followed by a reflection in the plane $y = x$.

$$M_1^2 = \begin{pmatrix} -1 & 0 & 0 \\ 0 & -1 & 0 \\ 0 & 0 & 1 \end{pmatrix}.$$

M_1^2 defines a rotation through 180° about Oz.
$M_2^2 = I$, since M_2 is its own inverse.

Transpose of a Matrix and Transpose of a Vector

Definitions 5.7 and 5.8 apply to three by three matrices, and Definition 5.9 may be extended to apply to a three-dimensional vector. The transpose of the vector $\mathbf{v} = \begin{pmatrix} x \\ y \\ z \end{pmatrix}$ is the row vector $\mathbf{v}' = (x \quad y \quad z)$.

The rule for computing products may also be extended.

If $M = \begin{pmatrix} a & b & c \\ d & e & f \\ g & h & q \end{pmatrix}$ and $\mathbf{v} = \begin{pmatrix} x \\ y \\ z \end{pmatrix}$, then $M\mathbf{v} = \begin{pmatrix} ax+by+cz \\ dx+ey+fz \\ gx+hy+qz \end{pmatrix}$,

and $\mathbf{v}'M = (ax+dy+gz \quad bx+ey+hz \quad cx+fy+qz)$.

It is clear that $I\mathbf{v} = \mathbf{v}$, and that $\mathbf{v}'\mathbf{v}_1$ is the inner product of the vectors $\mathbf{v}$ and $\mathbf{v}_1$.

It can be shown that if $\mathbf{v}_1 = M\mathbf{v}$, then $\mathbf{v}_1' = \mathbf{v}'M'$, and, provided M^{-1} exists, that $\mathbf{v} = M^{-1}\mathbf{v}_1$. The method of establishing these is similar to that used in Example (ix) of Chapter 5.

EXAMPLE (viii). Show that $\mathbf{v}_1'\mathbf{v}_2 = \mathbf{v}_2'\mathbf{v}_1$ for any three-dimensional vectors $\mathbf{v}_1$ and $\mathbf{v}_2$.

Let $$\mathbf{v}_1 = \begin{pmatrix} x_1 \\ y_1 \\ z_1 \end{pmatrix}, \quad \text{and} \quad \mathbf{v}_2 = \begin{pmatrix} x_2 \\ y_2 \\ z_2 \end{pmatrix}.$$

Then $$\mathbf{v}_1'\mathbf{v}_2 = (x_1 \quad y_1 \quad z_1)\begin{pmatrix} x_2 \\ y_2 \\ z_2 \end{pmatrix} = x_1x_2+y_1y_2+z_1z_2,$$

and $$\mathbf{v}_2'\mathbf{v}_1 = (x_2 \quad y_2 \quad z_2)\begin{pmatrix} x_1 \\ y_1 \\ z_1 \end{pmatrix} = x_2x_1+y_2y_1+z_2z_1.$$

Therefore $$\mathbf{v}_1'\mathbf{v}_2 = \mathbf{v}_2'\mathbf{v}_1.$$

Note. We could have tackled this question alternatively by stating that $\mathbf{v}_1'\mathbf{v}_2$ is the inner product $\mathbf{v}_1 \cdot \mathbf{v}_2$; since this operation is commutative, we deduce that $\mathbf{v}_1'\mathbf{v}_2 = \mathbf{v}_2'\mathbf{v}_1$.

EXAMPLE (ix). Show that for any three by three matrix M, the matrix $M'M$ is symmetric. Deduce that if M is orthogonal, then $M'M = I$.

Let the columns of M be denoted by the vectors $\mathbf{v}_1, \mathbf{v}_2, \mathbf{v}_3$, so that

$$M = (\mathbf{v}_1 \quad \mathbf{v}_2 \quad \mathbf{v}_3), \quad \text{and} \quad M' = \begin{pmatrix} \mathbf{v}_1' \\ \mathbf{v}_2' \\ \mathbf{v}_3' \end{pmatrix}.$$

Then
$$M'M = \begin{pmatrix} \mathbf{v}_1'\mathbf{v}_1 & \mathbf{v}_1'\mathbf{v}_2 & \mathbf{v}_1'\mathbf{v}_3 \\ \mathbf{v}_2'\mathbf{v}_1 & \mathbf{v}_2'\mathbf{v}_2 & \mathbf{v}_2'\mathbf{v}_3 \\ \mathbf{v}_3'\mathbf{v}_1 & \mathbf{v}_3'\mathbf{v}_2 & \mathbf{v}_3'\mathbf{v}_3 \end{pmatrix}.$$

The result of Example (viii) shows that this is symmetric. If M is orthogonal, then $\mathbf{v}_1$, $\mathbf{v}_2$ and $\mathbf{v}_3$ are perpendicular unit vectors. That is, $\mathbf{v}_1'\mathbf{v}_1 = \mathbf{v}_2'\mathbf{v}_2 = \mathbf{v}_3'\mathbf{v}_3 = 1$, and all the other entries of $M'M$, being the inner product of perpendicular vectors, are zero. Therefore $M'M = I$.

We conclude by solving Example (iv) again by a new method, replacing some of the traditional algebra in our former solution with some matrix algebra. The problem was to find the image of the surface S, whose equation is

$$41x^2+75y^2+34z^2-24xz = 25,$$

under the mapping defined by the orthogonal matrix

$$U = \tfrac{1}{5}\begin{pmatrix} 3 & 0 & 4 \\ 0 & 5 & 0 \\ 4 & 0 & -3 \end{pmatrix}.$$

The mapping is defined by the vector relation

$$U\mathbf{v} = \mathbf{v}_1, \quad \text{where} \quad \mathbf{v} = \begin{pmatrix} x \\ y \\ z \end{pmatrix}, \quad \text{and} \quad \mathbf{v}_1 = \begin{pmatrix} x_1 \\ y_1 \\ z_1 \end{pmatrix}.$$

It follows that $\mathbf{v} = U^{-1}\mathbf{v}_1$.

Since U is orthogonal, $U^{-1} = U'$ (established in Example (ix)),

and since U is symmetric, $U' = U$.

Therefore $\mathbf{v} = U\mathbf{v}_1$.

That is,
$$x = \tfrac{3}{5}x_1 + \tfrac{4}{5}z_1,$$
$$y = y_1,$$
$$z = \tfrac{4}{5}x_1 - \tfrac{3}{5}z_1.$$

The method continues from this point as in Example (iv).

Summary of Chapter 6

Mappings of x–y–z-space defined by three by three matrices are affine mappings. Under the mapping whose matrix is M, the volume enclosed by any surface is multiplied by $|\det M|$.

The operation of matrix multiplication may be extended to apply to three by three matrices and three-dimensional vectors.

Orthogonal three by three matrices define isometric mappings. Singular three by three matrices define many–one mappings. The inverse of a matrix M defines the inverse of the mapping defined by M. The product of two matrices defines the product of the mappings they each define.

If U is an orthogonal three by three matrix, then $U^{-1} = U'$, and $\det U = \pm 1$.

Exercise 6b

1. Find the matrix of the mapping which is equivalent to a reflection of x–y–z-space in the plane $y = 0$ followed by a reflection in the plane $y = x$,
(a) geometrically,
(b) by multiplying the appropriate matrices.

2. Prove that the product of two orthogonal matrices is an orthogonal matrix.

3. X is the singular matrix $\begin{pmatrix} 0 & 1 & -1 \\ 2 & 0 & 1 \\ 4 & 2 & 0 \end{pmatrix}$. What geometry mappings do X, X^2, X^3 define? Explain the apparent paradox of your answers.

4. D is the matrix $\begin{pmatrix} a & 0 & 0 \\ 0 & b & 0 \\ 0 & 0 & c \end{pmatrix}$. Find D^2, D^3, and deduce a formula for D^n, where n is a positive integer. Does your formula remain valid if n has the value -1?

5. M is the matrix $\begin{pmatrix} 1 & 2 & 3 \\ 4 & 5 & 6 \\ 7 & 8 & 9 \end{pmatrix}$, $\mathbf{v}$ the vector $\begin{pmatrix} x \\ y \\ z \end{pmatrix}$. Form, where possible, the products $M\mathbf{v}$, $M'\mathbf{v}$, $M\mathbf{v}'$, $M'\mathbf{v}'$, $\mathbf{v}'M$, $\mathbf{v}'M'$, $\mathbf{v}M$, $\mathbf{v}M'$, $M'M$, MM', $\mathbf{v}\mathbf{v}'$, $\mathbf{v}'\mathbf{v}$,

6. Show that multiplication of a three by three matrix by three-dimensional vectors is distributive over vector addition. That is, show that $M(\mathbf{v}_1+\mathbf{v}_2) = M\mathbf{v}_1+M\mathbf{v}_2$, where M is any three by three matrix and $\mathbf{v}_1$ and $\mathbf{v}_2$ are three-dimensional vectors.

7. For any three by three non-singular matrices M, N, show that

$$(MN)^{-1} = N^{-1}M^{-1}.$$

CHAPTER 7

GENERALISED VECTORS AND MATRICES

Introduction

In Chapter 1 we defined vectors as geometric quantities. We defined addition of vectors and multiplication of vectors by scalars, and we deduced rules for adding coordinate vectors and multiplying them by scalars.

But vector and matrix algebra can also be used to solve non-geometric problems. In the sciences, physical, biological and social, data are often presented in the form of an array suitable for treatment as a matrix. Linear relations exist between quantities, with no particular geometric significance. In this chapter we shall show how the concept of vectors and matrices may be modified to apply in such non-geometric contexts.

We shall therefore cease to demand a geometric significance for a vector quantity, and instead consider generalised coordinate vectors satisfying rules for addition and multiplication by a scalar. For such generalised vectors, we shall replace the word *coordinate* by the word *component*. We shall consider vectors with two, three, four, or n components, denoting a vector with n components by

$$\mathbf{v} = \begin{pmatrix} x \\ y \\ \vdots \\ t \end{pmatrix},$$

the dots signifying that not all the components of $\mathbf{v}$ are shown here.

n-dimensional Vectors

DEFINITION 7.1. *n-dimensional vectors are quantities specified by n components, such as*

$$\mathbf{v} = \begin{pmatrix} x \\ y \\ \vdots \\ t \end{pmatrix},$$

with properties of addition and multiplication by a scalar, as follows:

$$\begin{pmatrix} x_1 \\ y_1 \\ \vdots \\ t_1 \end{pmatrix} + \begin{pmatrix} x_2 \\ y_2 \\ \vdots \\ t_2 \end{pmatrix} = \begin{pmatrix} x_1+x_2 \\ y_1+y_2 \\ \vdots \\ t_1+t_2 \end{pmatrix}, \quad k\begin{pmatrix} x \\ y \\ \vdots \\ t \end{pmatrix} = \begin{pmatrix} kx \\ ky \\ \vdots \\ kt \end{pmatrix}.$$

Note. The rules for coordinate vectors, which were derived as consequences of the original definitions for geometric vectors, are now taken as definitions for component vectors.

EXAMPLE (i). A steel works produces two types of steel—tool steel and stainless steel. The constituents of each type are listed below. Derive a vector representing the quantities required of each constituent for an order of a tons of tool steel and b tons of stainless steel.

	Tool steel	*Stainless steel*
Pig iron	60%	70%
Chromium	10%	25%
Tungsten	18%	0%
Aluminium	0%	4%
Others	12%	1%

The constituents of 1 ton of tool steel may be represented by the components of the vector

$$\mathbf{v}_1 = \begin{pmatrix} 0{\cdot}60 \\ 0{\cdot}10 \\ 0{\cdot}18 \\ 0{\cdot}00 \\ 0{\cdot}12 \end{pmatrix},$$

each entry referring to the quantity required of one constituent.

A similar vector $\mathbf{v}_2$ may be constructed representing the constituents of 1 ton of stainless steel. Then the quantities required for the order of a tons of tool steel and b tons of stainless steel are given by the components of the vector

$$\mathbf{v} = a\mathbf{v}_1 + b\mathbf{v}_2 = \begin{pmatrix} 0{\cdot}6a + 0{\cdot}7b \\ 0{\cdot}1a + 0{\cdot}25b \\ 0{\cdot}18a + 0b \\ 0a + 0{\cdot}04b \\ 0{\cdot}12a + 0{\cdot}01b \end{pmatrix}.$$

DEFINITION 7.2. *The inner product of two n-dimensional vectors*

$$\mathbf{v} = \begin{pmatrix} x_1 \\ y_1 \\ \vdots \\ t_1 \end{pmatrix}, \quad \mathbf{v}_2 = \begin{pmatrix} x_2 \\ y_2 \\ \vdots \\ t_2 \end{pmatrix},$$

is the scalar quantity

$$x_1x_2 + y_1y_2 + \ldots + t_1t_2.$$

The quantity may be written as $\mathbf{v}_1 \cdot \mathbf{v}_2$, or as $\mathbf{v}_1'\mathbf{v}_2$.

EXAMPLE (ii). The cost per ton of the constituents of the two types of steel mentioned in Example (i) is given below. Derive an inner product of vectors giving the total cost of the order specified in Example (i).

Constituent	Pig iron	Chromium	Tungsten	Aluminium	Others
Cost per ton	£p	£c	£t	£a	£x

The vector $\mathbf{w} = \begin{pmatrix} p \\ c \\ t \\ a \\ x \end{pmatrix}$ has components which are the cost per ton of the constituents listed in the same order as they were for the vectors $\mathbf{v}_1$ and $\mathbf{v}_2$ of Example (i).

The total cost of the order will be

$$p(0{\cdot}6a+0{\cdot}7b)+c(0{\cdot}1a+0{\cdot}25b)+t(0{\cdot}18a)+a(0{\cdot}04b)+x(0{\cdot}12a+0{\cdot}01b).$$

This quantity is the inner product of vectors $\mathbf{w}'\mathbf{v}$, where $\mathbf{v} = a\mathbf{v}_1+b\mathbf{v}_2$, as in Example (i).

Polynomials as Vectors

The expression

$$p(t) = p_0+p_1t+p_2t^2+p_3t^3+\ldots+p_nt^n,$$

where n is a positive integer, and the dots signify that not all such similar terms of $p(t)$ have been written down, is called a polynomial in t of degree n.

We may represent the polynomial $p(t)$ by the vector

$$\mathbf{p} = \begin{pmatrix} p_0 \\ p_1 \\ \vdots \\ p_n \end{pmatrix}.$$

The sum of such vectors represents the sum of polynomials, and if we extend the rule for multiplying a matrix by a vector, we shall see a significance of this operation in Example (iii).

EXAMPLE (iii). Find a matrix which maps the polynomial

$$p(t) = p_0+p_1t+p_2t^2+p_3t^3$$

onto its derivative with respect to t, $p'(t)$.

If you know nothing of the theory of differentiation, you should omit this example. If you do, you will know that

$$p'(t) = p_1+2p_2t+3p_3t^2.$$

Thus if $p(t)$ is represented by the vector $\begin{pmatrix} p_0 \\ p_1 \\ p_2 \\ p_3 \end{pmatrix}$, $p'(t)$ will be represented by the vector $\begin{pmatrix} p_1 \\ 2p_2 \\ 3p_3 \\ 0 \end{pmatrix}$.

Now $$\begin{pmatrix} 0 & 1 & 0 & 0 \\ 0 & 0 & 2 & 0 \\ 0 & 0 & 0 & 3 \\ 0 & 0 & 0 & 0 \end{pmatrix} \begin{pmatrix} p_0 \\ p_1 \\ p_2 \\ p_3 \end{pmatrix} = \begin{pmatrix} p_1 \\ 2p_2 \\ 3p_3 \\ 0 \end{pmatrix}.$$

That is, the above matrix maps $p(t)$ onto $p'(t)$.

Age Distribution Vectors

In some sociological or biological studies it is convenient to list the members of a population according to age groups. For instance, a rabbit population of a certain area, containing i_0 immature rabbits and a_0 adult rabbits (*adult* signifying over one year old and therefore of breeding age), may be represented by the vector $\mathbf{v}_0 = \begin{pmatrix} i_0 \\ a_0 \end{pmatrix}$. The sum of such vectors has obvious significance. Multiplication of a matrix by such a vector can have interesting significance, as is shown in Example (iv).

EXAMPLE (iv). A rabbit population increases at the rate of four live infants per adult rabbit per year. If one-third of the whole population survives every year, derive a vector equation giving the relation between the population one year and the population the subsequent year.

Let $\mathbf{v}_0 = \begin{pmatrix} i_0 \\ a_0 \end{pmatrix}$ and $\mathbf{v}_1 = \begin{pmatrix} i_1 \\ a_1 \end{pmatrix}$ represent the populations of two consecutive years. From the data of the question, we deduce that

$$i_1 = 4a_0,$$

$$a_1 = \tfrac{1}{3}i_0 + \tfrac{1}{3}a_0.$$

That is, $\mathbf{v}_1 = M\mathbf{v}_0$, where $M = \begin{pmatrix} 0 & 4 \\ \frac{1}{3} & \frac{1}{3} \end{pmatrix}$.

Note. The matrix M of Example (iv) is called a transition matrix. It provides a means of predicting the rabbit population of the area for any subsequent year. For instance, if $\mathbf{v}_2$ represents the population after two years, then

$$\mathbf{v}_2 = M\mathbf{v}_1.$$

But, since $\mathbf{v}_1 = M\mathbf{v}_0$,

$$\mathbf{v}_2 = M^2\mathbf{v}_0.$$

Similarly, $\mathbf{v}_n = M^n\mathbf{v}_0$,

where $\mathbf{v}_n$ represents the population after n years.

Exercise 7a

1. The following sets of quantities may be represented by generalised vectors. For each case, decide what significance, if any, might be attached to the operations (a) vector addition (or subtraction), (b) multiplication by a scalar, (c) multiplication by a matrix.

(i) The prices of certain commodities in a given grocery store on various days.

(ii) The age, height and weight of various people.

(iii) The ingredients of various cakes.

(iv) The national export figures, for various months, of machinery, cars, chemicals and electrical goods.

2. A baby food is to be manufactured from the ingredients carrot, cereal, chicken and flavouring. The vitamin contents of these ingredients, in mg per ounce, are as follows:

	Vitamin A	*Thiamine*	*Riboflavin*
Carrot	1	0·02	0·01
Cereal	0	0·10	0·05
Chicken	0	0·02	0·05
Flavouring	0	0	0

Arrange a suitable matrix–vector product giving in vector form the vitamin content of a mixture of x parts carrot, y parts cereal and z parts chicken.

3. Represent the polynomial $p(t) = p_0+p_1t+p_2t^2+p_3t^3$ as a vector. Find a matrix which maps this vector onto that representing $p(2t) = p_0+p_1(2t)+p_2(2t)^2+p_3(2t)^3$. Find also the matrix mapping it onto that representing $p(\frac{1}{2}t)$. Calculate the product of the two matrices, and explain your answer.

4. In Example (iii) of the text we found a matrix M mapping the polynomial $p(t)$ onto its derivative $p'(t)$. Calculate M^2 and explain what mapping this matrix defines. Deduce a value of n for which $M^n = 0$, the zero matrix.

5. With the transition matrix for rabbits given in Example (iv) of the text, show that a population with three times as many immature as adult rabbits can be expected to remain in the same proportion every year, but to increase by $33\frac{1}{3}\%$ per year.

6. A certain type of beetle has a life span of three years. It propagates in its third year of life, each beetle producing an average six live offspring. Half the beetles survive their first year of life, and one-third of the remainder survive their second year. At a certain time there are x_0 beetles in the first year of life, y_0 in the second and z_0 in the third. If

$$\mathbf{v}_0 = \begin{pmatrix} x_0 \\ y_0 \\ z_0 \end{pmatrix},$$

and $\mathbf{v}_1$ is a similar vector representing the population one year later, find a transition matrix such that

$$M\mathbf{v}_0 = \mathbf{v}_1.$$

Rectangular Matrices

In earlier examples of this chapter, we extended the rules for matrix multiplication to apply to vectors in more than three dimensions and to four by four matrices. We shall now show that significance may be attached to matrices with any number of rows or columns, even if there is not the same number of rows as columns.

DEFINITION 7.3. *A matrix is an array of numbers in rows and columns, with properties of multiplication and addition, as follows:*

(1) *The product MN of the matrices M, N exists provided the rows of M and the columns of N have the same number of entries. In this case suppose the vectors $\mathbf{r}'_1, \mathbf{r}'_2, \ldots, \mathbf{r}'_m$ represent the rows of M, and that the vectors $\mathbf{c}_1, \mathbf{c}_2, \ldots, \mathbf{c}_n$ represent the columns of N.*

Then
$$MN = \begin{pmatrix} \mathbf{r}'_1\mathbf{c}_1 & \mathbf{r}'_1\mathbf{c}_2 & \ldots & \mathbf{r}'_1\mathbf{c}_n \\ \mathbf{r}'_2\mathbf{c}_1 & \mathbf{r}'_2\mathbf{c}_2 & \ldots & \mathbf{r}'_2\mathbf{c}_n \\ \cdot & \cdot & \ldots & \cdot \\ \mathbf{r}'_m\mathbf{c}_1 & \mathbf{r}'_m\mathbf{c}_2 & \ldots & \mathbf{r}'_m\mathbf{c}_n \end{pmatrix}.$$

(2) *The sum $M+N$ of the matrices M, N exists provided M and N have the same number of rows and the same number of columns. In this case $M+N$ is the matrix whose entries are the sums of the corresponding entries of M and N.*

In Definition 7.3 we have introduced a new operation, matrix addition. Notice that this operation is equivalent to vector addition for the case of a matrix with one column or one row.

EXAMPLE (v). The matrix $M = \begin{pmatrix} 1 & 2 & 3 \\ 4 & 5 & 6 \end{pmatrix}$, and the matrix $N = \begin{pmatrix} a & b \\ c & d \\ e & f \end{pmatrix}$. Compute, if they exist, the following: MN, NM, $M'N$, $N'M$, $M+N$, $M'+N$, $M+N'$.

The rows of M have three entries, as do the columns of N. Therefore the product MN exists.

$$MN = \begin{pmatrix} 1 & 2 & 3 \\ 4 & 5 & 6 \end{pmatrix} \begin{pmatrix} a & b \\ c & d \\ e & f \end{pmatrix} = \begin{pmatrix} a+2c+3e & b+2d+3f \\ 4a+5c+6e & 4b+5d+6f \end{pmatrix},$$

$$NM = \begin{pmatrix} a & b \\ c & d \\ e & f \end{pmatrix} \begin{pmatrix} 1 & 2 & 3 \\ 4 & 5 & 6 \end{pmatrix} = \begin{pmatrix} a+4b & 2a+5b & 3a+6b \\ c+4d & 2c+5d & 3c+6d \\ e+4f & 2e+5f & 3e+6f \end{pmatrix}.$$

The rows of M' have two entries, the columns of N three.

Therefore the product $M'N$ does not exist.

Similarly, the product $N'M$ does not exist.

M and N do not have the same number of rows or the same number of columns. Therefore the sum $M+N$ does not exist.

M' and N have the same number of rows and the same number of columns.

$$M'+N = \begin{pmatrix} 1 & 4 \\ 2 & 5 \\ 3 & 6 \end{pmatrix} + \begin{pmatrix} a & b \\ c & d \\ e & f \end{pmatrix} = \begin{pmatrix} a+1 & b+4 \\ c+2 & d+5 \\ e+3 & f+6 \end{pmatrix},$$

$$M+N' = \begin{pmatrix} 1 & 2 & 3 \\ 4 & 5 & 6 \end{pmatrix} + \begin{pmatrix} a & c & e \\ b & d & f \end{pmatrix} = \begin{pmatrix} a+1 & c+2 & e+3 \\ b+4 & d+5 & f+6 \end{pmatrix}.$$

Note. A matrix with the same number of rows as columns is called a *square* matrix. A matrix without this property is called *rectangular.*

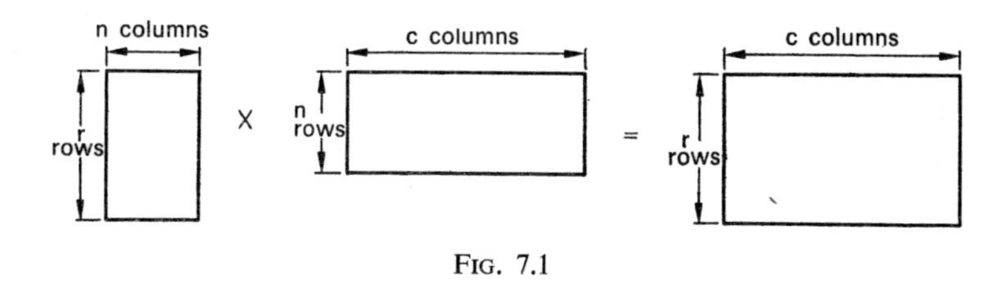

FIG. 7.1

In Fig. 7.1 we illustrate the "shape" of the product of two rectangular matrices, an r by n matrix multiplied by an n by c matrix.

Incidence Matrices

Matrices may have significance in themselves without being associated with a mapping. We now introduce such an example. Figure 7.2 illustrates three railway routes, 1, 2 and 3, which run between five towns, A, B, C, D and E.

Town A lies on lines 1 and 3 but not on line 2. This is recorded in the matrix below by writing in the column headed A the number 1

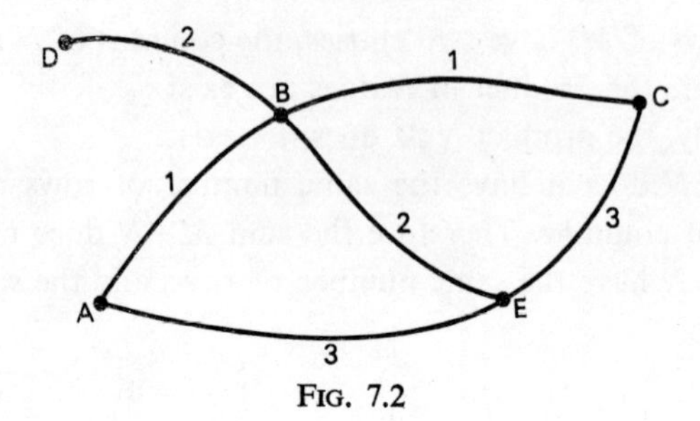

FIG. 7.2

in the first and third rows, and the number 0 in the second row.

	A	B	C	D	E
1	1	1	1	0	0
2	0	1	0	1	1
3	1	0	1	0	1

Such a matrix is called an *incidence matrix*.

EXAMPLE (vi). If M is the incidence matrix associated with Fig. 7.2, calculate and interpret the significance of the matrices $M'M$, MM'.

$$M'M = \begin{pmatrix} 2 & 1 & 2 & 0 & 1 \\ 1 & 2 & 1 & 1 & 1 \\ 2 & 1 & 2 & 0 & 1 \\ 0 & 1 & 0 & 1 & 1 \\ 1 & 1 & 1 & 1 & 2 \end{pmatrix}.$$

Notice that $M'M$ is a symmetric matrix. Its first entry, 2, signifies that town A lies on two routes; the second entry of the first row, 1, signifies that towns A and B are connected by one route; the third entry of this row signifies that towns A and C are connected by two routes. Every entry gives the number of routes connecting some pair of towns.

$$MM' = \begin{pmatrix} 3 & 1 & 2 \\ 1 & 3 & 1 \\ 2 & 1 & 3 \end{pmatrix}.$$

MM' is also a symmetric matrix. Its first entry, 3, signifies that Route 1 serves three stations; the second entry, 1, signifies that Routes 1 and 2 meet in one station; the third entry, 2, signifies that Routes 1 and 3 meet in two stations. Every entry has a corresponding significance.

Dominance Matrices

Consider a football league of four teams which all play one another in matches where no draws are allowed. The results may be illustrated as in Fig. 7.3, where the arrows point in each case towards the loser. (For instance, Team 1 beat Teams 2 and 3, but lost to Team 4.)

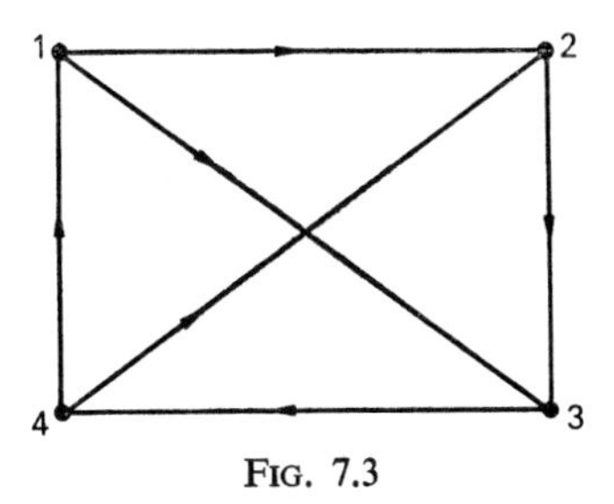

FIG. 7.3

The results may also be tabulated in the form of a matrix, where a win for any team is recorded by writing the number 1 in the row allotted to that team, and a loss by writing 0. The diagonal entries of this matrix, representing the result of no match, are each filled with 0.

	1	2	3	4
1	0	1	1	0
2	0	0	1	0
3	0	0	0	1
4	1	1	0	0

Such a matrix is called a dominance matrix. Team 2 is said to be dominant to Team 3, and Team 3 dominant to Team 4. That Team 2 is not dominant to Team 4 should not seem incongruous. (If you know what a transitive relation is, you will recognise that dominance is not a transitive relation.)

First scrutiny of the matrix suggests that Teams 1 and 4 should be rated as equal winners, and Teams 2 and 3 as equal losers. But suppose we need to select a winning team. It might be suggested that Team 4's win over Team 1 is worth more than Team 1's win over Team 3, and that therefore Team 4 merits the top position.

One mathematical way of establishing a winning team is to consider indirect dominance. Team 2, having beaten Team 3, which in its turn beat Team 4, is said to be indirectly dominant to Team 4. Notice that Team 4, having beaten Team 1, which in its turn beat Team 2, is thus indirectly dominant to Team 2. If we compute M^2, where M is the original dominance matrix, we find that its entries give the number of ways in which each team is indirectly dominant to the others. (For instance, Team 4 is indirectly dominant to Team 3 in two ways, via Teams 1 and 2.)

$$M^2 = \begin{pmatrix} 0 & 0 & 1 & 1 \\ 0 & 0 & 0 & 1 \\ 1 & 1 & 0 & 0 \\ 0 & 1 & 2 & 0 \end{pmatrix}.$$

EXAMPLE (vii). M is the dominance matrix associated with Fig. 7.3. Compute the matrix $M+M^2$, and derive a list of the teams in order of superiority, based on direct and indirect dominance.

$$M+M^2 = \begin{pmatrix} 0 & 1 & 2 & 1 \\ 0 & 0 & 1 & 1 \\ 1 & 1 & 0 & 1 \\ 1 & 2 & 2 & 0 \end{pmatrix}.$$

The entries of the matrix $M+M^2$ give the number of ways each team is dominant, directly or indirectly, to every other team. (For instance, Team 1 dominates Team 3 twice, once directly and once indirectly.)

Adding the entries of each row of the matrix $M+M^2$, we obtain the total number of dominances for each team. Thus Team 1 has four, Team 2 has two, Team 3 has three and Team 4 has five. This suggests that the order of superiority based on dominance is Team 4, Team 1, Team 3, Team 2.

Summary of Chapter 7

We defined n-dimensional non-geometric, or generalised, vectors to obey certain rules for addition and for multiplication by a scalar, and also rules for inner product, when this has significance.

We extended the concept of a matrix to cover arrays with any number of rows or columns, and formulated rules of multiplication and addition of such matrices.

As examples of non-geometric or generalised vectors, we met constituent vectors, polynomial vectors, and age-distribution vectors.

As examples of non-geometric or generalised matrices, we met transition matrices, incidence matrices and dominance matrices.

Exercise 7b

1. M is the matrix $\begin{pmatrix} 1 & 2 \\ 3 & 4 \\ 5 & 6 \end{pmatrix}$, N the matrix $\begin{pmatrix} a & b \\ c & d \\ e & f \end{pmatrix}$, and P the matrix $\begin{pmatrix} x & y \\ z & t \end{pmatrix}$.

Show that

$$MP+NP = (M+N)P.$$

What property of matrix operations have you illustrated?

2. Set up an incidence matrix M corresponding to the points and lines in Fig. 7.4. Calculate the product matrices $M'M$, MM', and interpret the significance of each.

3. Set up a dominance matrix corresponding to Fig. 7.5, which illustrates the pecking habits of five chickens. (For instance, Chicken 1 pecks Chickens 2, 3 and 4, but is pecked by Chicken 5.) Determine the "top chicken", and also a "pecking order" for the other chickens.

4. The matrix M below is the dominance matrix for seven entrants in a "round-robin" tennis tournament. Show that the rows of the matrix $M+M^2$ do not reveal a winning entrant. Interpret the entries of M^3, and show that the rows of

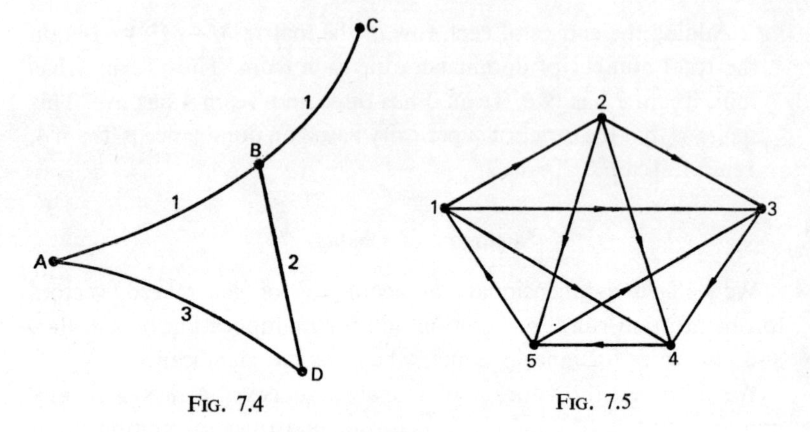

FIG. 7.4

FIG. 7.5

the matrix $M+M^2+M^3$ reveal a winner of the tournament.

$$\begin{pmatrix} 0 & 0 & 0 & 1 & 1 & 1 & 1 \\ 1 & 0 & 1 & 0 & 0 & 0 & 1 \\ 1 & 0 & 0 & 0 & 1 & 1 & 1 \\ 0 & 1 & 1 & 0 & 1 & 1 & 0 \\ 0 & 1 & 0 & 0 & 0 & 0 & 0 \\ 0 & 1 & 0 & 0 & 1 & 0 & 0 \\ 0 & 0 & 0 & 1 & 1 & 1 & 0 \end{pmatrix}.$$

CHAPTER 8

LINEAR EQUATIONS

Introduction

The variables in a set of linear equations may be considered as components of a generalised vector. For example, the quantities x, y and z in the equations

$$\begin{aligned} 2x+2y-z &= 7, \\ -x+2y+2z &= -5, \\ 2x-y+2z &= 4 \end{aligned}$$

may be considered as components of a three-dimensional vector $\mathbf{v}$, and the equations may be expressed in the form

$$M\mathbf{v} = \mathbf{v}_1,$$

where $$M = \begin{pmatrix} 2 & 2 & -1 \\ -1 & 2 & 2 \\ 2 & -1 & 2 \end{pmatrix}, \quad \text{and} \quad \mathbf{v}_1 = \begin{pmatrix} 7 \\ -5 \\ 4 \end{pmatrix}.$$

EXAMPLE (i). Solve the simultaneous equations

$$\begin{aligned} 2x+2y-z &= 7, \\ -x+2y+2z &= -5, \\ 2x-y+2z &= 4. \end{aligned}$$

We have already expressed these equations in the form $M\mathbf{v} = \mathbf{v}_1$.

Now, $\frac{1}{3}M$ is an orthogonal matrix. Putting $\frac{1}{3}M = U$, we may rewrite this vector equation

$$3U\mathbf{v} = \mathbf{v}_1,$$

or

$$U\mathbf{v} = \tfrac{1}{3}\mathbf{v}_1.$$

Applying the technique we established in Example (ix) of Chapter 5, we see that

$$\mathbf{v} = U^{-1}\tfrac{1}{3}\mathbf{v}_1.$$

Now, since U is orthogonal, $U^{-1} = U'$.

$$\text{Therefore } \mathbf{v} = \frac{1}{3}\begin{pmatrix} 2 & -1 & 2 \\ 2 & 2 & -1 \\ -1 & 2 & 2 \end{pmatrix} \frac{1}{3}\begin{pmatrix} 7 \\ -5 \\ 4 \end{pmatrix} = \begin{pmatrix} 3 \\ 0 \\ -1 \end{pmatrix}.$$

Therefore the solution to the equations is $x = 3$, $y = 0$, $z = -1$.

To use this technique for solving linear equations we shall need a method for finding the inverse of any square matrix. We shall establish such a method when we have introduced the concept of elementary row operations.

Elementary Row Operations

DEFINITION 8.1. *Elementary row operations are operations which change the entries of one or more rows of a matrix in one or more of the following ways:*

(a) *replace each entry of a row by k times itself (k being some real, non-zero number),*

(b) *replace each entry of a row by itself added to k times the corresponding entry of another row (k being some real, non-zero number),*

(c) *interchange two rows.*

A row operation on a matrix M can be effected by multiplying M on the left by another matrix, called an *elementary matrix*. An elementary matrix is one resulting from some elementary row operation on I, the identity matrix with the same number of rows as M.

For example, doubling the first row of $I = \begin{pmatrix} 1 & 0 \\ 0 & 1 \end{pmatrix}$, we obtain the matrix $E_1 = \begin{pmatrix} 2 & 0 \\ 0 & 1 \end{pmatrix}$. Then, if $M = \begin{pmatrix} a & b & c \\ d & e & f \end{pmatrix}$, then

$$E_1M = \begin{pmatrix} 2a & 2b & 2c \\ d & e & f \end{pmatrix}.$$

That is, E_1M is the matrix resulting from the same row operation on M.

Again, replacing the second row of I by itself added to twice the first row, we obtain the matrix $E_2 = \begin{pmatrix} 1 & 0 \\ 2 & 1 \end{pmatrix}$.

$$E_2M = \begin{pmatrix} a & b & c \\ d+2a & e+2b & f+2c \end{pmatrix}.$$

That is, E_2M is the matrix resulting from the same row operation on M.

EXAMPLE (ii). Find elementary matrices which multiply successively on the left of $M = \begin{pmatrix} 2 & -1 \\ 1 & 3 \end{pmatrix}$ to result in I.

We shall try to choose elementary row operations which change the entries of M to those of I. To obtain 1 as the first entry, we might interchange the two rows of M. The interchange of these rows will be effected by multiplying by the elementary matrix

$$E_1 = \begin{pmatrix} 0 & 1 \\ 1 & 0 \end{pmatrix}.$$

$$E_1M = \begin{pmatrix} 1 & 3 \\ 2 & -1 \end{pmatrix}.$$

Next, to obtain 0 instead of 2 in the matrix E_1M, we could replace the second row of this matrix by itself added to -2 times the first row. This is effected by multiplying by the elementary matrix

$$E_2 = \begin{pmatrix} 1 & 0 \\ -2 & 1 \end{pmatrix}.$$

$$E_2E_1M = \begin{pmatrix} 1 & 3 \\ 0 & -7 \end{pmatrix}.$$

Next, we shall replace the entry -7 by 1. This is effected by multiplying the matrix E_2E_1M by the elementary matrix

$$E_3 = \begin{pmatrix} 1 & 0 \\ 0 & -\frac{1}{7} \end{pmatrix}.$$

$$E_3E_2E_1M = \begin{pmatrix} 1 & 3 \\ 0 & 1 \end{pmatrix}.$$

Lastly, we multiply by the elementary matrix

$$E_4 = \begin{pmatrix} 1 & -3 \\ 0 & 1 \end{pmatrix}.$$

$$E_4E_3E_2E_1M = I.$$

Note. The matrix product $E_4E_3E_2E_1$ must be the inverse of M. To find M^{-1}, we might calculate this product; alternatively, we could perform the same series of row operations on the matrix I as we did on M. Doing this, we should obtain successively the matrices E_1I, E_2E_1I, $E_3E_2E_1I$, and finally the matrix $E_4E_3E_2E_1I$, which is the inverse of M.

Inverse of a Square Matrix

EXAMPLE (iii). Calculate the inverse of the matrix $M = \begin{pmatrix} 2 & -1 \\ 1 & 3 \end{pmatrix}$ by elementary row operations.

We shall perform elementary row operations changing the entries of M to those of I, and perform the same operations on I, which will change its entries to those of M^{-1}. Although we shall record the successive matrices as the products of other matrices, we shall actually obtain them by performing elementary row operations. To record each row operation, we shall use the notation $\mathbf{r}_1^*$ to mean "new first row", and $\mathbf{r}_1$ to mean "previous first row". Thus the first row operation, recorded by the equations $\mathbf{r}_1^* = \mathbf{r}_2, \mathbf{r}_2^* = \mathbf{r}_1$, is the interchange of rows. (For simplicity we omit the usual notation $\mathbf{r}_1'$ for a row vector.)

$$M = \begin{pmatrix} 2 & -1 \\ 1 & 3 \end{pmatrix} \qquad\qquad I = \begin{pmatrix} 1 & 0 \\ 0 & 1 \end{pmatrix}$$

$$E_1M = \begin{pmatrix} 1 & 3 \\ 2 & -1 \end{pmatrix} \quad \left.\begin{matrix} \mathbf{r}_1^* = \mathbf{r}_2, \\ \mathbf{r}_2^* = \mathbf{r}_1. \end{matrix}\right\} \qquad E_1I = \begin{pmatrix} 0 & 1 \\ 1 & 0 \end{pmatrix}$$

$$E_2E_1M = \begin{pmatrix} 1 & 3 \\ 0 & -7 \end{pmatrix} \quad \mathbf{r}_2^* = \mathbf{r}_2 - 2\mathbf{r}_1 \qquad E_2E_1I = \begin{pmatrix} 0 & 1 \\ 1 & -2 \end{pmatrix}$$

$$E_3E_2E_1M = \begin{pmatrix} 1 & 3 \\ 0 & 1 \end{pmatrix} \quad \mathbf{r}_2^* = -\tfrac{1}{7}\mathbf{r}_2 \qquad E_3E_2E_1I = \begin{pmatrix} 0 & 1 \\ -\frac{1}{7} & \frac{2}{7} \end{pmatrix}$$

$$E_4E_3E_2E_1M = \begin{pmatrix} 1 & 0 \\ 0 & 1 \end{pmatrix} \quad \mathbf{r}_1^* = \mathbf{r}_1 - 3\mathbf{r}_2 \qquad E_4E_3E_2E_1I = \begin{pmatrix} \frac{3}{7} & \frac{1}{7} \\ -\frac{1}{7} & \frac{2}{7} \end{pmatrix}$$

Hence $\qquad M^{-1} = E_4E_3E_2E_1I = \frac{1}{7}\begin{pmatrix} 3 & 1 \\ -1 & 2 \end{pmatrix}.$

Check that $MM^{-1} = I$.

An alternative way of setting out a calculation is offered in Example (iv), where M and I are combined into one rectangular matrix. No reference is made in this example to the elementary matrices, which are not necessary for the computation but are the basis for the justification of the method.

The new symbol used in Example (iv) denotes *row equivalence*. The relation $M \sim N$ denotes that N may be obtained by performing some elementary row operations on M; we say that M and N are *row equivalent*.

EXAMPLE (iv). Compute the inverse of the matrix $M = \begin{pmatrix} 2 & 4 & -1 \\ 1 & 3 & 1 \\ -1 & 0 & 3 \end{pmatrix}$.

$$\begin{pmatrix} 2 & 4 & -1 & 1 & 0 & 0 \\ 1 & 3 & 1 & 0 & 1 & 0 \\ -1 & 0 & 3 & 0 & 0 & 1 \end{pmatrix}$$

$$\sim \begin{pmatrix} 1 & 3 & 1 & 0 & 1 & 0 \\ 2 & 4 & -1 & 1 & 0 & 0 \\ -1 & 0 & 3 & 0 & 0 & 1 \end{pmatrix} \quad \left.\begin{aligned} \mathbf{r}_1^* &= \mathbf{r}_2 \\ \mathbf{r}_2^* &= \mathbf{r}_1 \end{aligned}\right\}$$

$$\sim \begin{pmatrix} 1 & 3 & 1 & 0 & 1 & 0 \\ 0 & -2 & -3 & 1 & -2 & 0 \\ 0 & 3 & 4 & 0 & 1 & 1 \end{pmatrix} \quad \begin{aligned} \mathbf{r}_2^* &= \mathbf{r}_2 - 2\mathbf{r}_1 \\ \mathbf{r}_3^* &= \mathbf{r}_3 + \mathbf{r}_1 \end{aligned}$$

$$\sim \begin{pmatrix} 1 & 3 & 1 & 0 & 1 & 0 \\ 0 & 1 & 1 & 1 & -1 & 1 \\ 0 & 3 & 4 & 0 & 1 & 1 \end{pmatrix} \quad \mathbf{r}_2^* = \mathbf{r}_2 + \mathbf{r}_3$$

$$\sim \begin{pmatrix} 1 & 0 & -3 & 0 & 0 & -1 \\ 0 & 1 & 1 & 1 & -1 & 1 \\ 0 & 0 & 1 & -3 & 4 & -2 \end{pmatrix} \quad \begin{aligned} \mathbf{r}_1^* &= \mathbf{r}_1 - 3\mathbf{r}_2 \\ \mathbf{r}_3^* &= \mathbf{r}_3 - 3\mathbf{r}_2 \end{aligned}$$

$$\sim \begin{pmatrix} 1 & 0 & 0 & -9 & 12 & -7 \\ 0 & 1 & 0 & 4 & -5 & 3 \\ 0 & 0 & 1 & -3 & 4 & -2 \end{pmatrix} \quad \begin{aligned} \mathbf{r}_1^* &= \mathbf{r}_1 + 3\mathbf{r}_3 \\ \mathbf{r}_2^* &= \mathbf{r}_2 - \mathbf{r}_3 \end{aligned}$$

Hence $$M^{-1} = \begin{pmatrix} -9 & 12 & -7 \\ 4 & -5 & 3 \\ -3 & 4 & -2 \end{pmatrix}.$$

Check that $M^{-1}M = I$.

Note. In this example we have sometimes left out a step, writing down the result of two successive row operations. When taking short cuts of this nature, make quite sure they are valid.

A singular matrix is not row equivalent to I. If we try to carry out the process of reducing a singular matrix to I by row operations, we shall at some stage obtain a complete row of zeros. For example, the matrix $M = \begin{pmatrix} 1 & -1 & 1 \\ 2 & 3 & 7 \\ 3 & 2 & 8 \end{pmatrix}$ is singular, for the sum of its first and second rows is the third row. So if we perform on M two successive row operations symbolised by $\mathbf{r}_1^* = \mathbf{r}_1 + \mathbf{r}_2 - \mathbf{r}_3$, we obtain

$$M \sim \begin{pmatrix} 0 & 0 & 0 \\ 2 & 3 & -1 \\ 1 & 2 & 2 \end{pmatrix}.$$

However we attempt to reduce the matrix M to I, we shall at some stage reach a situation similar to this.

Elementary Column Operations

Multiplication on the right by elementary matrices results in elementary operations on columns instead of on rows.

For example, $$\begin{pmatrix} a & b \\ c & d \end{pmatrix} \begin{pmatrix} 2 & 0 \\ 0 & 1 \end{pmatrix} = \begin{pmatrix} 2a & c \\ 2b & d \end{pmatrix}.$$

Elementary column operations could be used to determine the inverse of a square matrix. For instance, if

$$ME_1E_2E_3 \ldots E_n = I,$$

then

$$E_1E_2E_3 \ldots E_n = M^{-1}.$$

But a mixture of row and column operations will not enable us to find M^{-1}. If, for instance,

$$E_2E_1ME_3E_4 = I,$$

we can deduce nothing useful about M^{-1}.

Exercise 8a

1. Find the elementary matrices E_a, E_b, E_c, which multiply on the left of any four by four matrix to:

(a) interchange its first and fourth rows,
(b) replace its third row by 5 times itself,
(c) replace its third row by itself added to twice the second row.

2. Compute by elementary row operations the inverses, if they exist, of the following matrices:

$$\text{(a)}\ \begin{pmatrix} 1 & 2 \\ 5 & 3 \end{pmatrix}, \quad \text{(b)}\ \begin{pmatrix} 0 & 1 & 2 \\ -1 & 3 & 0 \\ 1 & -2 & 1 \end{pmatrix}, \quad \text{(c)}\ \begin{pmatrix} 2 & 3 & -1 \\ 4 & -1 & -9 \\ 1 & 2 & 0 \end{pmatrix},$$

$$\text{(d)}\ \begin{pmatrix} 1 & 0 & 2 & -1 \\ 0 & 1 & 1 & 1 \\ -1 & 1 & 2 & 1 \\ 0 & -1 & 1 & -1 \end{pmatrix}, \quad \text{(e)}\ \begin{pmatrix} 1 & 0 & 0 \\ 0{\cdot}02 & 0{\cdot}10 & 0{\cdot}02 \\ 0{\cdot}01 & 0{\cdot}05 & 0{\cdot}05 \end{pmatrix}.$$

3. The matrix $M = \begin{pmatrix} 1 & 2 & 3 \\ 3 & 3 & 3 \\ 4 & 5 & 6 \end{pmatrix}$. Find a matrix T, such that $TM = \begin{pmatrix} 1 & 2 & 3 \\ 5 & 7 & 9 \\ 6 & 6 & 6 \end{pmatrix}$.

4. Use the results of question 2 to solve the following sets of linear equations:

$$\text{(a)}\quad \begin{aligned} y+2z &= -12, \\ -x+3y \quad &= 15, \\ x-2y+z &= -20; \end{aligned} \qquad \text{(b)}\quad \begin{aligned} p-q-2r-s &= -7, \\ p \quad +2r-s &= 5, \\ q+r+s &= -3, \\ q-r+s &= 1. \end{aligned}$$

5. Question 2 of Exercise 7a gives the vitamin content of certain ingredients of a baby food. Calculate in what proportion these ingredients should be mixed in order to produce a mixture with the three vitamins in the ratio 5 : 4 : 3. Part (e) of question 2 will help.

Consistent and Inconsistent Equations

Not every set of linear equations is soluble by the matrix method outlined in the first part of this chapter. Consider, for example, the equations

$$\begin{aligned} x+\ y &= 2, \\ 2x+2y &= 4. \end{aligned}$$

These equations may be expressed in the form $M\mathbf{v} = \mathbf{v}_1$, where

$$\mathbf{v} = \begin{pmatrix} x \\ y \end{pmatrix}, \quad \mathbf{v}_1 = \begin{pmatrix} 2 \\ 4 \end{pmatrix}, \quad \text{and} \quad M = \begin{pmatrix} 1 & 1 \\ 2 & 2 \end{pmatrix}.$$

But since M is a singular matrix, we cannot find a solution to this vector equation in the form $\mathbf{v} = M^{-1}\mathbf{v}_1$.

In fact, the equations have infinitely many solutions. The value of x may be arbitrarily chosen and then the value of y satisfying $x+y=2$ will satisfy both equations. The general solution may be expressed as

$$x \text{ arbitrary}, \ y = 2-x,$$

or as

$$y \text{ arbitrary}, \ x = 2-y.$$

Next, consider the equations

$$\begin{aligned} x+y &= 2, \\ 2x+2y &= 5. \end{aligned}$$

If we express these in the form $M\mathbf{v} = \mathbf{v}_1$, the matrix M is the same singular matrix as appeared in the previous illustration. So again, we cannot find a solution of the form $\mathbf{v} = M^{-1}\mathbf{v}_1$.

In fact, since these two equations contradict each other, there are no solutions for x and y satisfying both. We call such equations *inconsistent*.

We shall now outline a method of solution which reveals all possible solutions of any set of linear equations.

The Echelon Form

An *echelon matrix* is one which has the properties

(i) the first non-zero entry of every row is 1, and

(ii) in each successive row this entry of 1 occurs progressively further to the right.

Examples of echelon matrices include

$$\begin{pmatrix} 1 & 3 & 7 & 9 \\ 0 & 1 & 6 & 5 \\ 0 & 0 & 0 & 1 \\ 0 & 0 & 0 & 0 \end{pmatrix}, \quad \begin{pmatrix} 0 & 1 & 0 & 5 \\ 0 & 0 & 1 & 2 \end{pmatrix}, \quad \begin{pmatrix} 1 & 3 \\ 0 & 1 \\ 0 & 0 \end{pmatrix}, \quad \begin{pmatrix} 1 & 0 \\ 0 & 1 \end{pmatrix}.$$

The word *echelon* means *rung of a ladder*, and we can think of an echelon matrix as composed of a ladder of 1's, supported by zeros.

Every matrix is row equivalent to some echelon matrix. We shall see how this fact provides a method of solving any set of linear equations.

EXAMPLE (v). Solve the simultaneous equations

$$\begin{aligned} x+2y-\ z &= \ \ 0, \\ 2x \qquad -8z &= \ \ 6, \\ -x-y+3z &= -1. \end{aligned}$$

We form a matrix M from the coefficients on the left sides of the equations and the numbers on the right sides. This matrix will be called the augmented matrix of the equations.

$$M = \begin{pmatrix} 1 & 2 & -1 & 0 \\ 2 & 0 & -8 & 6 \\ -1 & -1 & 3 & -1 \end{pmatrix}.$$

The solution of any set of equations is unaltered if:

(a) every term on both sides of one equation is multiplied by any non-zero number k,

(b) every term on both sides of one equation is replaced by itself added to k times the corresponding term of another equation, or
(c) any two of the equations are interchanged.

These three cases correspond to the performing of elementary row operations on the augmented matrix M. For instance, replacing the third row of M by itself added to the first row gives

$$M \sim \begin{pmatrix} 1 & 2 & -1 & 0 \\ 2 & 0 & -8 & 6 \\ 0 & 1 & 2 & -1 \end{pmatrix}.$$

The third row of this matrix implies the equation

$$y+2z = -1,$$

an equation obtainable by adding the first and third equations of the original set.

We continue performing elementary row operations on M until we have an echelon matrix.

$$M \sim \begin{pmatrix} 1 & 2 & -1 & 0 \\ 0 & -4 & -6 & 6 \\ 0 & 1 & 2 & -1 \end{pmatrix} \qquad \mathbf{r}_2^* = \mathbf{r}_2 - 2\mathbf{r}_1$$

$$\sim \begin{pmatrix} 1 & 2 & -1 & 0 \\ 0 & 1 & 2 & -1 \\ 0 & 0 & 1 & 1 \end{pmatrix} \qquad \begin{aligned} \mathbf{r}_2^* &= \mathbf{r}_3 \\ \mathbf{r}_3^* &= \tfrac{1}{2}\mathbf{r}_2 + 2\mathbf{r}_3. \end{aligned}$$

We may read off the solutions to the equations by "climbing up the ladder" of the equations implied by the echelon matrix, thus:

$$\begin{aligned} z &= 1 && \text{(implied by the third row),} \\ y &= -1-2z = -3 && \text{(implied by the second row),} \\ x &= z-2y = 7 && \text{(implied by the first row).} \end{aligned}$$

EXAMPLE (vi). Solve for x, y, z the equations:

$$\begin{aligned} x-\ y-3z &= 3, \\ 2x-3y-\ z &= 5, \\ x-2y+2z &= 2. \end{aligned}$$

Let $$M = \begin{pmatrix} 1 & -1 & -3 & 3 \\ 2 & -3 & -1 & 5 \\ 1 & -2 & 2 & 2 \end{pmatrix}.$$

$$M \sim \begin{pmatrix} 1 & -1 & -3 & 3 \\ 0 & -1 & 5 & -1 \\ 0 & -1 & 5 & -1 \end{pmatrix} \qquad \begin{aligned} \mathbf{r}_2^* &= \mathbf{r}_2 - 2\mathbf{r}_1 \\ \mathbf{r}_3^* &= \mathbf{r}_3 - \mathbf{r}_1 \end{aligned}$$

$$\sim \begin{pmatrix} 1 & -1 & -3 & 3 \\ 0 & 1 & -5 & 1 \\ 0 & 0 & 0 & 0 \end{pmatrix} \qquad \begin{aligned} \mathbf{r}_2^* &= -\mathbf{r}_2 \\ \mathbf{r}_3^* &= \mathbf{r}_3 - \mathbf{r}_2. \end{aligned}$$

The echelon matrix has a final row implying the equation $0 = 0$. The second row implies the equation $y = 1+5z$, from which we deduce that the value of z may be arbitrarily chosen. The general solution may be expressed as

$$\begin{aligned} &z \text{ arbitrary}, \\ &y = 1+5z, \\ &x = 3+3z+y = 4+8z. \end{aligned}$$

Note. In examples (v) and (vi) we have reduced the augmented matrix to an echelon matrix in a systematic manner; we first obtained 1 as the first entry of the first row, and then obtained a column of zeros below it; next we obtained 1 as the second entry of the second row, and then obtained a column of zeros below it. If we were using a computer to solve a set of linear equations, we should program the computer to perform elementary row operations on the augmented matrix in the same systematic order. The computer would print out the final echelon matrix, from which we would read off the solutions.

EXAMPLE (vii). Solve for x, y and z the equations

$$\begin{aligned} x+2y-\ z &= \ \ 0, \\ 2x \qquad -8z &= \ \ 6, \\ -x-\ y+3z &= -1, \\ x+\ y-\ z &= \ \ 4. \end{aligned}$$

$$\text{Let } M = \begin{pmatrix} 1 & 2 & -1 & 0 \\ 2 & 0 & -8 & 6 \\ -1 & -1 & 3 & -1 \\ 1 & 1 & -1 & 4 \end{pmatrix}.$$

$$M \sim \begin{pmatrix} 1 & 2 & -1 & 0 \\ 0 & 1 & 2 & -1 \\ 0 & 0 & 1 & 1 \\ 0 & 0 & 2 & 3 \end{pmatrix} \quad \mathbf{r}_4^* = \mathbf{r}_4+\mathbf{r}_3, \text{ other operations as in Example (v).}$$

$$\sim \begin{pmatrix} 1 & 2 & -1 & 0 \\ 0 & 1 & 2 & -1 \\ 0 & 0 & 1 & 1 \\ 0 & 0 & 0 & 1 \end{pmatrix}.$$

The final row of the echelon matrix implies the equation $0 = 1$. This being false implies that the equations are inconsistent.

EXAMPLE (viii). Solve for p, q, r and s the equations

$$\begin{aligned} p+\ q-\ r+s &= 2, \\ 3p \qquad +2r \qquad &= 3, \\ 2p-\ q+3r-s &= 1, \\ 4p+\ q+\ r+s &= 5. \end{aligned}$$

$$\text{Let} \qquad M = \begin{pmatrix} 1 & 1 & -1 & 1 & 2 \\ 3 & 0 & 2 & 0 & 3 \\ 2 & -1 & 3 & -1 & 1 \\ 4 & 1 & 1 & 1 & 5 \end{pmatrix}.$$

$$M \sim \begin{pmatrix} 1 & 1 & -1 & 1 & 2 \\ 0 & 1 & -\frac{5}{3} & 1 & 1 \\ 0 & 0 & 0 & 0 & 0 \\ 0 & 0 & 0 & 0 & 0 \end{pmatrix}.$$

The echelon matrix implies the solutions

$$\begin{aligned} &s \text{ arbitrary},\\ &r \text{ arbitrary},\\ &q = 1-s+\tfrac{5}{3}r,\\ &p = 2+r-s-q = 1-\tfrac{2}{3}r. \end{aligned}$$

EXAMPLE (ix). Find values of p and q for which the following equations have (a) an infinite number of solutions, (b) no solution:

$$\begin{aligned} x+y+z &= 5,\\ -x+3y+2z &= p,\\ -2x-y+qz &= 0. \end{aligned}$$

Let $$M = \begin{pmatrix} 1 & 1 & 1 & 5\\ -1 & 3 & 2 & p\\ -2 & -1 & q & 0 \end{pmatrix}.$$

Check that $$M \sim \begin{pmatrix} 1 & 1 & 1 & 5\\ 0 & 1 & q+2 & 10\\ 0 & 0 & -4q-5 & p-35 \end{pmatrix}.$$

(a) The above is an echelon matrix implying an infinite number of solutions to the equations, provided its final row consists of zeros.

That is, $$-4q-5 = p-35 = 0,$$
giving $$q = -\tfrac{5}{4},\quad p = 35.$$

(b) If there are no solutions to the equations, then the final row of the above matrix must consist of three zeros and a fourth non-zero number.

That is, $$-4q-5 = 0,\quad p-35 \neq 0,$$
giving $$q = -\tfrac{5}{4},\quad p \neq 35.$$

EXAMPLE (x). Show that the matrix $M = \begin{pmatrix} 2 & 3 & -1\\ 4 & -1 & -9\\ 1 & 2 & 0 \end{pmatrix}$ is singular.

In Chapter 6 we noted that if the columns of a square matrix form linearly dependent vectors, then the matrix is singular.

The columns of M are linearly dependent vectors if numbers a, b and c exist, not all zero, such that

$$a\begin{pmatrix}2\\4\\1\end{pmatrix}+b\begin{pmatrix}3\\-1\\2\end{pmatrix}+c\begin{pmatrix}-1\\-9\\0\end{pmatrix}=\begin{pmatrix}0\\0\\0\end{pmatrix}.$$

That is, a, b and c satisfy a set of linear equations whose augmented matrix is

$$M_1=\begin{pmatrix}2 & 3 & -1 & 0\\4 & -1 & -9 & 0\\1 & 2 & 0 & 0\end{pmatrix}.$$

$$M_1\sim\begin{pmatrix}1 & 2 & 0 & 0\\0 & -7 & -7 & 0\\0 & -1 & -1 & 0\end{pmatrix}$$

$$\sim\begin{pmatrix}1 & 2 & 0 & 0\\0 & 1 & 1 & 0\\0 & 0 & 0 & 0\end{pmatrix}.$$

The echelon matrix implies an infinite number of non-zero solutions for a, b and c. Therefore the columns of M are linearly dependent vectors, and the matrix is singular.

Note. The working for Example (x) is almost identical to that we should employ to show that M is not row-equivalent to I. Showing this would also imply that M is singular. Compare the working too to your attempt, in question 2 part (c) of Exercise 8a, to compute the inverse of the same matrix.

EXAMPLE (xi). $\mathbf{e}_1$, $\mathbf{e}_2$ are two linearly independent two-dimensional vectors. Show that any other two-dimensional vector $\mathbf{v}$ may be expressed in the form $\mathbf{v} = a\mathbf{e}_1+b\mathbf{e}_2$, where a and b are real numbers.

Let $$\mathbf{v}=\begin{pmatrix}x\\y\end{pmatrix},\quad \mathbf{e}_1=\begin{pmatrix}x_1\\y_1\end{pmatrix},\quad \mathbf{e}_2=\begin{pmatrix}x_2\\y_2\end{pmatrix}.$$

We require then to show that the equations

$$ax_1+bx_2 = x,$$
$$ay_1+by_2 = y$$

have solutions for a and b.

The equations may be written in the form

$$M\begin{pmatrix}a\\b\end{pmatrix} = \mathbf{v}, \quad \text{where} \quad M = (\mathbf{e}_1, \mathbf{e}_2).$$

Since $\mathbf{e}_1$, $\mathbf{e}_2$ are linearly independent, M is a non-singular matrix, so it has an inverse, M^{-1}.

Therefore the equations have a solution $\begin{pmatrix}a\\b\end{pmatrix} = M^{-1}\mathbf{v}$.

Note. We may deduce from the result of Example (xi) that any three two-dimensional vectors are linearly dependent, a fact we illustrated without proof in Chapter 3.

Calculating Aids

Solving a set of linear equations can be a tedious process when the number of equations is large, when the number of variables is large, or when the coefficients contain a large number of digits. For instance, to solve the equations

$$0{\cdot}4096a+0{\cdot}1654b+0{\cdot}0872c+0{\cdot}1010d = 0{\cdot}4043,$$
$$0{\cdot}8194a-0{\cdot}0304b-0{\cdot}9164c+0{\cdot}8189d = 0{\cdot}1550,$$
$$0{\cdot}2557a-1{\cdot}7102b+0{\cdot}0068c-0{\cdot}5375d = 0{\cdot}4240,$$

you would probably resort to a calculating aid, such as log tables, slide-rule, desk calculator or computer. With each aid you would need to consider the accuracy to which you required your solution, and the effect of "rounding-off" numbers during the process of solution. Such "rounding off" can cause considerable changes to the solutions, and in fact render them useless. Consider, for example, the equations

$$0{\cdot}21x+0{\cdot}30y = 0{\cdot}1,$$
$$0{\cdot}29x+0{\cdot}42y = 0{\cdot}1.$$

The solution to these is $x = 10$, $y = -\frac{20}{3}$.

But the equations

$$0{\cdot}2x+0{\cdot}3y = 0{\cdot}1,$$
$$0{\cdot}3x+0{\cdot}4y = 0{\cdot}1,$$

obtained by "rounding off" the coefficients in the original set to one decimal place, have as solution $x = -1$, $y = 1$.

This set of equations is an example of ill-conditioned equations. If we wrote such equations in the form

$$M\mathbf{v} = \mathbf{v}_1,$$

where M is a square matrix, we should find that the matrix M is almost singular—that is, that det M is small compared with the entrie of M.

Summary of Chapter 8

We introduced the unknowns in a set of linear equations as components of a generalised vector $\mathbf{v}$, expressing the equations in the form

$$M\mathbf{v} = \mathbf{v}_1,$$

where M is a matrix and $\mathbf{v}_1$ the vector of constants.

We defined elementary row operations on a matrix, and derived a method of calculating, where it exists, the inverse of a square matrix, by means of such operations. We solved some sets of linear equations by using the inverse of a matrix.

Sets of linear equations may be consistent or inconsistent. If consistent, they have either one unique solution or an infinite number of solutions. We solved generally sets of linear equations by using elementary row operations to reduce an augmented matrix to an echelon matrix.

Exercise 8b

1. By reducing an augmented matrix to echelon form, solve generally the following sets of equations:

(a)
$$\begin{aligned} x-\ y-\ z &= 1,\\ 2x-3y-4z &= 1,\\ 2x+\ y+4z &= 5. \end{aligned}$$

(b)
$$\begin{aligned} x-\ y+\ z &= 2,\\ 2x-3y+\ z &= 0,\\ 2x+\ y+5z &= 1. \end{aligned}$$

(c)
$$\begin{aligned} a-2b+\ c &= 9,\\ -3a+\ b+7c &= -2,\\ 2a+6b-\ c &= 2,\\ a-3b-3c &= 2. \end{aligned}$$

(d)
$$\begin{aligned} p-2q-3r+3s &= 0,\\ 2p-3q-\ r+5s &= 4,\\ -p+\ q+2r-4s &= 3. \end{aligned}$$

2. Find the conditions on a, b and c so that the following sets of equations have solutions:

(a)
$$\begin{aligned} -4x+3y+az &= b,\\ 5x-4y+cz &= a. \end{aligned}$$

(b)
$$\begin{aligned} 4p-3q &= a,\\ -3p+2q &= b,\\ 5p-4q &= c. \end{aligned}$$

3. Question 2 of Exercise 7a concerns the preparation of a baby food. A new variety is now to be produced, with spinach replacing the cereal in the former. Spinach contains 0·5 mg vitamin A per ounce, 0·02 mg thiamine and 0·03 mg riboflavin. Find all possible proportions of the ingredients which will produce a mixture with the vitamins in the ratio 25 : 2 : 4.

4. M is the matrix $\begin{pmatrix} 1 & 1 & -3\\ 2 & 3 & -1\\ 1 & 2 & 2 \end{pmatrix}$. Show that M is singular, by:

(a) showing that its rows are linearly dependent (this should be obvious),
(b) showing that its columns are linearly dependent,
(c) showing that M is not row-equivalent to I, and
(d) evaluating det M.

5. Show that the three planes, whose equations are given by

$$\begin{aligned} 8x-5y+3z &= 11,\\ 3x+4y-\ z &= 2,\\ 5x-9y+4z &= 9, \end{aligned}$$

have a line in common.

6. Show that there are three values of p for which the equations

$$\begin{aligned} x-py\qquad &= 0,\\ 4x-3y+pz &= 0,\\ px+\ y+2z &= 0 \end{aligned}$$

have a non-trivial solution, and find these values. Find the solution for x, y and z in terms of p, assuming that p takes one of these values.

7. $\mathbf{e}_1$, $\mathbf{e}_2$ and $\mathbf{e}_3$ are linearly independent three-dimensional vectors. Show that any other vector $\mathbf{v}$ may be expressed in the form

$$\mathbf{v} = a\mathbf{e}_1 + b\mathbf{e}_2 + c\mathbf{e}_3,$$

where a, b and c are real numbers. Draw a conclusion concerning any four three-dimensional vectors.

8. Solve for x, y and z the following sets of equations:

(a)
$$\begin{aligned} x+\tfrac{1}{2}y+\tfrac{1}{3}z &= 1, \\ \tfrac{1}{2}x+\tfrac{1}{3}y+\tfrac{1}{4}z &= \tfrac{3}{4}, \\ \tfrac{1}{3}x+\tfrac{1}{4}y+\tfrac{1}{5}z &= \tfrac{11}{20}. \end{aligned}$$

(b)
$$\begin{aligned} x+0{\cdot}50y+0{\cdot}33z &= 1, \\ 0{\cdot}5x+0{\cdot}33y+0{\cdot}25z &= 0{\cdot}75, \\ 0{\cdot}33x+0{\cdot}25y+0{\cdot}20z &= 0{\cdot}55. \end{aligned}$$

Explain the difference in the two sets of solutions.

9. Find the condition for the equations of Example (ix) to have a unique solution.

CHAPTER 9

EIGENVECTORS

Introduction

We met the concept of *invariance* in Chapter 4, when we discussed mappings. In this chapter we shall investigate a particular invariant property of some mappings. *Eigen* is a German word meaning special; and an eigenvector of a mapping is a special vector whose direction is invariant under, or reversed by, the mapping.

For example, the mapping of the x–y-plane whose matrix is $\begin{pmatrix} 2 & 0 \\ 0 & 3 \end{pmatrix}$ maps the unit square onto a rectangle, as illustrated in Fig. 9.1.

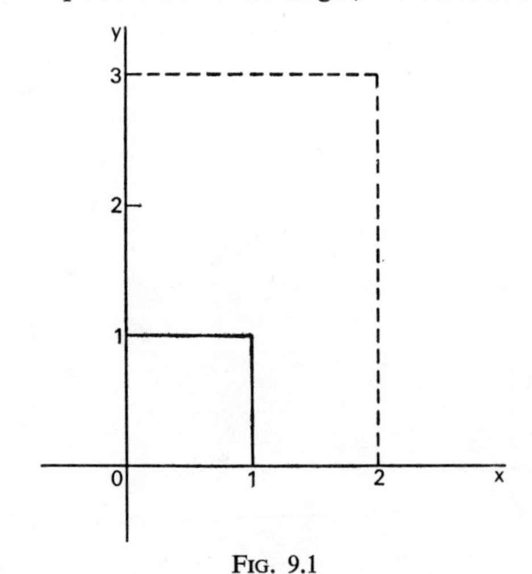

FIG. 9.1

Clearly, the image of the vector **i** under this mapping is 2**i**. Thus, **i** is a vector whose direction is unchanged by the mapping—that is, **i** is an eigenvector of the mapping. Another eigenvector is **j**. But, taking some vector at random, say the vector $\begin{pmatrix}1\\1\end{pmatrix}$, we find that its image, $\begin{pmatrix}2\\3\end{pmatrix}$, has a different direction from it. This illustrates that we can find vectors which are certainly not eigenvectors.

Our discussion has now reverted to geometric significance. In this chapter we shall initially concern ourselves with mappings of the x–y-plane and of x–y–z-space. Then, generalising our results, we shall find we have some useful tools for solving non-geometric problems.

Eigenvectors of a Matrix

DEFINITION 9.1. *Given a square matrix M, if* **e** *is a non-zero vector for which*

$$M\mathbf{e} = k\mathbf{e}, \textit{ for some real number } k,$$

then **e** *is an eigenvector of the matrix M, and k is an eigenvalue of M corresponding to* **e**.

EXAMPLE (i). Show that the vector $\begin{pmatrix}1\\1\end{pmatrix}$ is an eigenvector of the matrix $M = \begin{pmatrix}4 & -1\\2 & 1\end{pmatrix}$. Find the eigenvalue corresponding to this vector.

Under the mapping defined by M, the image of the vector $\begin{pmatrix}1\\1\end{pmatrix}$ is given by

$$\begin{pmatrix}4 & -1\\2 & 1\end{pmatrix}\begin{pmatrix}1\\1\end{pmatrix} = \begin{pmatrix}3\\3\end{pmatrix}.$$

Thus the image of $\begin{pmatrix}1\\1\end{pmatrix}$ is $\begin{pmatrix}3\\3\end{pmatrix}$, a vector in the same direction.

Thus $\begin{pmatrix}1\\1\end{pmatrix}$ is an eigenvector of M, and the corresponding eigenvalue is 3.

Figure 9.2 shows the image of the unit square under this mapping, illustrating that the direction of $\begin{pmatrix}1\\1\end{pmatrix}$ is the same as that of its image.

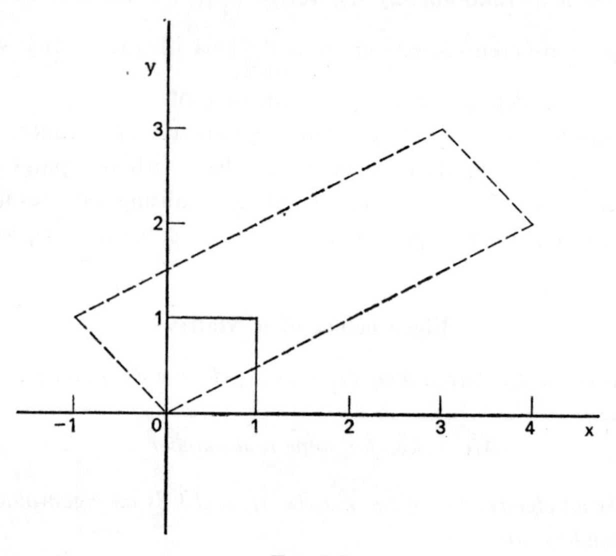

FIG. 9.2

Does every square matrix have eigenvectors? How can we find all the eigenvectors of a matrix, and how many can we expect to find? Let us try to discover all the eigenvectors of the matrix M of Example (i).

EXAMPLE (ii). Find all the values of x, y and k which satisfy the equation $M\mathbf{e} = k\mathbf{e}$, where $\mathbf{e} = \begin{pmatrix}x\\y\end{pmatrix}$, and $M = \begin{pmatrix}4 & -1\\2 & 1\end{pmatrix}$.

The equation $M\mathbf{e} = k\mathbf{e}$ implies the linear equations

$$4x-y = kx,$$
$$2x+y = ky.$$

These may be rearranged as

$$\begin{aligned}(4-k)\,x-y &= 0,\\ 2x+(1-k)y &= 0.\end{aligned}$$

The augmented matrix of these equations is

$$N = \begin{pmatrix} 4-k & -1 & 0 \\ 2 & 1-k & 0 \end{pmatrix}.$$

$$N \sim \begin{pmatrix} 1 & \dfrac{1-k}{2} & 0 \\ 0 & -1-\dfrac{(4-k)(1-k)}{2} & 0 \end{pmatrix} \quad \begin{aligned} &\mathbf{r}_1^* = \tfrac{1}{2}\mathbf{r}_2 \\ &\mathbf{r}_2^* = \mathbf{r}_1 - \frac{(4-k)}{2}\mathbf{r}_2. \end{aligned}$$

For the equations to have non-zero solutions,

$$-1-\frac{(4-k)(1-k)}{2} = 0.$$

That is, $$k^2-5k+6 = 0,$$

$$k = 3 \text{ or } 2.$$

For either of these values of k, the solution for x and y is

$$y \text{ arbitrary,} \quad x = \frac{k-1}{2}y.$$

That is, for $k = 3, \quad x = y,$

and for $k = 2, \quad x = \frac{1}{2}y.$

Thus we have two sets of eigenvectors of the matrix M: one set, $a\begin{pmatrix}1\\1\end{pmatrix}$, for any real non-zero number a, all parallel, corresponding to an eigenvalue of 3, and another set, $b\begin{pmatrix}1\\2\end{pmatrix}$, for any real non-zero number b, all parallel, corresponding to an eigenvalue of 2.

The Characteristic Equation

We shall now prove a theorem which will give us an alternative method for finding eigenvalues to that employed in Example (ii).

THEOREM 9.1. *If k is an eigenvalue of the square matrix M, then the matrix $(M-kI)$ is singular.*

Proof. Suppose $\mathbf{e}$ is an eigenvector of the matrix M, corresponding to the eigenvalue k.

Then $M\mathbf{e} = k\mathbf{e}$

$$= kI\mathbf{e}, \quad \text{since} \quad I\mathbf{e} = \mathbf{e}.$$

Subtracting the vector $kI\mathbf{e}$ from both sides of this vector equation,

$$M\mathbf{e} - kI\mathbf{e} = \mathbf{0}, \quad \text{the zero vector.}$$

Using the distributive property of matrix multiplication over matrix addition, we have

$$(M-kI)\mathbf{e} = \mathbf{0}.$$

Now, if the matrix $(M-kI)$ has an inverse, $(M-kI)^{-1}$, then

$$\begin{aligned} \mathbf{e} &= (M-kI)^{-1}\mathbf{0} \\ &= \mathbf{0}, \end{aligned}$$

so that $\mathbf{e}$ is a non-zero vector only if the matrix $(M-kI)$ has no inverse.

Thus the matrix $(M-kI)$ must be singular.

Note. It can be proved that conversely, if the matrix $(M-kI)$ is singular, then k is an eigenvalue of M.

Now we can use an alternative method of solving Example (ii).

If $M = \begin{pmatrix} 4 & -1 \\ 2 & 1 \end{pmatrix}$, then $(M-kI) = \begin{pmatrix} 4-k & -1 \\ 2 & 1-k \end{pmatrix}$.

If $(M-kI)$ is singular, then $\det(M-kI) = 0.$
That is, $(4-k)(1-k)+2 = 0,$
yielding two possible values of k, $k = 3$, or $k = 2$.

Now we look for an eigenvector $\mathbf{e}_1 = \begin{pmatrix} x_1 \\ y_1 \end{pmatrix}$, satisfying $(M-3I)\mathbf{e}_1 = \mathbf{0}$.

Then

$$\begin{aligned} x_1 - y_1 &= 0, \\ 2x_1 - 2y_1 &= 0. \end{aligned}$$

The solution of these equations is y_1 arbitrary, $x_1 = y_1$.

Similarly, the vector $\mathbf{e}_2 = \begin{pmatrix} x_2 \\ y_2 \end{pmatrix}$ satisfies $(M-2I)\mathbf{e}_2 = \mathbf{0}$ if y_2 is arbitrary, and $x_2 = \frac{1}{2}y_2$.

Note. The equation $\det(M-kI) = 0$, from which we derive values of k, is called the *characteristic equation* of M.

EXAMPLE (iii). Calculate the eigenvalues and eigenvectors of the matrix $\begin{pmatrix} 0 & 4 \\ \frac{1}{3} & \frac{1}{3} \end{pmatrix}$, the transition matrix for the rabbit population described in Example (iv) of Chapter 7. Interpret the results in terms of the rabbit population.

The characteristic equation of M is

$$-k(\tfrac{1}{3}-k)-\tfrac{4}{3} = 0,$$

which is satisfied by $k = \frac{4}{3}$ or -1.

The vector $\mathbf{e}_1 = \begin{pmatrix} x_1 \\ y_1 \end{pmatrix}$ satisfies $(M-\frac{4}{3})\mathbf{e}_1 = \mathbf{0}$ if

$$\begin{aligned} -\tfrac{4}{3}x_1 + 4y_1 &= 0, \\ \tfrac{1}{3}x_1 - y_1 &= 0, \end{aligned}$$

that is, if y_1 is arbitrary and $x_1 = 3y_1$.

Thus there is a set of eigenvectors $a\mathbf{e}_1$, where $\mathbf{e}_1 = \begin{pmatrix} 3 \\ 1 \end{pmatrix}$, for all real non-zero a, corresponding to the eigenvalue of $\frac{4}{3}$. This result may be applied to a rabbit population with an initial age distribution in the ratio 3:1. If this distribution is represented by the vector $\mathbf{e} = \begin{pmatrix} 3a \\ a \end{pmatrix}$, then, since $M\mathbf{e} = \frac{4}{3}\mathbf{e}$, we can expect the age distribution to remain in the same ratio and the population to increase by $33\frac{1}{3}\%$ every year.

Check that there is a set of eigenvectors $b\mathbf{e}_2$ where $\mathbf{e}_2 = \begin{pmatrix} 4 \\ -1 \end{pmatrix}$, corresponding to the eigenvalue of -1. There is no useful interpretation to be drawn from this result, since every vector in the set $b\mathbf{e}_2$ must contain a negative component.

EXAMPLE (iv). Show that the matrix $M = \begin{pmatrix} 2 & -4 \\ 1 & -1 \end{pmatrix}$ has no real eigenvalues.

The characteristic equation of M is

$$(2-k)(-1-k)+4 = 0.$$

That is, $$k^2+k+2 = 0.$$

The solution of this quadratic equation is

$$k = \frac{-1 \pm \sqrt{(1-8)}}{2}.$$

Since $\sqrt{(1-8)}$ is not a real number, M has no real eigenvalues.

EXAMPLE (v). Calculate the eigenvalues and eigenvectors of the matrix $M = \begin{pmatrix} 3 & 0 & 0 \\ 0 & 3 & -1 \\ 0 & 2 & 0 \end{pmatrix}$.

The matrix $(M-kI) = \begin{pmatrix} 3-k & 0 & 0 \\ 0 & 3-k & -1 \\ 0 & 2 & -k \end{pmatrix}$,

so that $$\det(M-kI) = (3-k)[-k(3-k)+2].$$

The characteristic equation of M is therefore

$$(3-k)(k^2-3k+2) = 0,$$

which is satisfied by $k = 1, 2$ or 3.

The vector $\mathbf{e}_1 = \begin{pmatrix} x_1 \\ y_1 \\ z_1 \end{pmatrix}$ satisfies $(M-I)\mathbf{e}_1 = \mathbf{0}$ if

$$\begin{aligned} 2x_1 \qquad &= 0, \\ 2y_1 - z_1 &= 0, \\ 2y_1 - z_1 &= 0, \end{aligned}$$

that is, if z_1 is arbitrary, $\quad y_1 = \frac{1}{2}z_1, \quad x_1 = 0.$

Thus we have a set of eigenvectors $a\begin{pmatrix} 0 \\ 1 \\ 2 \end{pmatrix}$, for real non-zero a, all parallel, corresponding to an eigenvalue of 1.

Check that there is another set of eigenvectors, $b\begin{pmatrix} 0 \\ 1 \\ 1 \end{pmatrix}$, corresponding to an eigenvalue of 2, and a third set, $c\begin{pmatrix} 1 \\ 0 \\ 0 \end{pmatrix}$, corresponding to an eigenvalue of 3.

Example (vi). Find the characteristic equation of the matrix $M = \begin{pmatrix} a & b \\ c & d \end{pmatrix}$. Deduce (a) that if M is singular, one of its eigenvalues is zero, and (b) that if M is symmetric and $b \neq 0$, its eigenvalues are real and distinct.

The characteristic equation of M is

$$(a-k)(c-k)-bc = 0,$$

that is, $\quad k^2-(a+d)k+ad-bc = 0.$

(a) If M is singular, then $\det M = ad-bc = 0$, so that the characteristic equation reduces to

$$k^2-(a+d)k = 0,$$

which has for one of its solutions $k = 0$.

(b) If M is symmetric, $b = c$. The solutions of the characteristic equation are then

$$k = \tfrac{1}{2}[a+d \pm \sqrt{\{(a+d)^2 - 4(ad-b^2)\}}]$$
$$= \tfrac{1}{2}[a+d \pm \sqrt{\{(a-d)^2 + 4b^2\}}].$$

The number inside this square root sign, being the sum of squares, must always be positive. (It cannot be zero, since $b \neq 0$.)

Therefore there will be two distinct (that is, different) and real values for k.

Note. Examples (iii) to (vi) illustrate that an n by n matrix has as its characteristic equation a polynomial equation of degree n. This equation yields up to n, but not more than n, eigenvalues.

Exercise 9a

1. Show that **i**, **j** and **k** are eigenvectors of the matrix $\begin{pmatrix} a & 0 & 0 \\ 0 & b & 0 \\ 0 & 0 & c \end{pmatrix}$.

What are the corresponding eigenvalues?

2. Verify that $\begin{pmatrix} 1 \\ 1 \end{pmatrix}$ and $\begin{pmatrix} -1 \\ 1 \end{pmatrix}$ are eigenvectors of the matrix $\begin{pmatrix} 3 & 1 \\ 1 & 3 \end{pmatrix}$. What are the corresponding eigenvalues?

3. We have defined the term eigenvector with reference to a square matrix. Why is it meaningless to talk of an eigenvector of a rectangular matrix?

4. Calculate the real eigenvalues and eigenvectors of the following matrices:

(a) $\begin{pmatrix} 4 & -1 \\ 1 & 2 \end{pmatrix}$, (b) $\begin{pmatrix} 2 & 1 \\ -1 & 1 \end{pmatrix}$, (c) $\begin{pmatrix} 0 & 0 & 6 \\ \frac{1}{2} & 0 & 0 \\ 0 & \frac{1}{3} & 0 \end{pmatrix}$, (d) $\begin{pmatrix} 2 & 0 & 0 \\ 0 & 2 & 2 \\ 0 & 2 & -1 \end{pmatrix}$?

5. Find relations between a, b, c and d if the matrix $\begin{pmatrix} a & b \\ c & d \end{pmatrix}$ has

(a) no real eigenvalues,
(b) two coincident eigenvalues.
Verify that these relations hold for the relevant matrices of question 4.

6. Verify that one of the eigenvalues of the singular matrix $\begin{pmatrix} 2 & 3 & -1 \\ 4 & -1 & -9 \\ 1 & 2 & 0 \end{pmatrix}$ is zero.

Calculate the corresponding set of eigenvectors.

7. The matrix of question 4(c) is the transition matrix for the beetle population described in question 6 of Exercise 7a. Interpret the results in terms of the beetle population.

8. The matrix $M = \begin{pmatrix} 4 & -1 \\ 2 & 1 \end{pmatrix}$, and $C = \begin{pmatrix} 1 & 1 \\ 1 & 2 \end{pmatrix}$. Compute the product $C^{-1}MC$. Refer to Example (ii) of this chapter to note where the eigenvectors and eigenvalues of M appear in this calculation.

9. The matrix $M = \begin{pmatrix} 3 & 0 & 0 \\ 0 & 3 & -1 \\ 0 & 2 & 0 \end{pmatrix}$, and $C = \begin{pmatrix} 1 & 0 & 0 \\ 0 & 1 & 1 \\ 0 & 1 & 2 \end{pmatrix}$. Compute the product $C^{-1}MC$. Refer to Example (v) of this chapter to note where the eigenvectors and eigenvalues of M appear in this calculation.

Diagonalisation

In this section we shall refer to questions 8 and 9 of Exercise 9a, so it is advisable to have attempted these.

DEFINITION 9.2. *A diagonal matrix is a square matrix whose only non-zero entries are in the leading diagonal of the matrix. (This consists of the first term of the first row, the second term of the second row, etc.)*

Examples of diagonal matrices include the identity I, the matrix $\begin{pmatrix} 3 & 0 \\ 0 & 2 \end{pmatrix}$, which you probably obtained as the product matrix $C^{-1}MC$ in question 8 of Exercise 9a, and $\begin{pmatrix} 3 & 0 & 0 \\ 0 & 2 & 0 \\ 0 & 0 & 1 \end{pmatrix}$, which you probably obtained as the product matrix $C^{-1}MC$ in question 9.

From answering questions 7 of Exercise 5b and 4 of Exercise 6b, you will have seen that it is particularly simple to compute powers of diagonal matrices. For example, if $D = \begin{pmatrix} a & 0 \\ 0 & b \end{pmatrix}$, then $D^n = \begin{pmatrix} a^n & 0 \\ 0 & b^n \end{pmatrix}$ for any positive integer n.

EXAMPLE (vii). If $C^{-1}MC = D$, show that $M^5 = CD^5C^{-1}$. Hence calculate M^5, where $M = \begin{pmatrix} 4 & -1 \\ 2 & 1 \end{pmatrix}$.

$$\begin{aligned} \text{If} \quad C^{-1}MC &= D, \\ \text{then} \quad CC^{-1}MCC^{-1} &= CDC^{-1}, \\ \text{that is,} \quad M &= CDC^{-1}. \\ \text{Thus} \quad M^5 &= CDC^{-1}CDC^{-1}CDC^{-1}CDC^{-1}CDC^{-1} \\ &= CD^5C^{-1}. \end{aligned}$$

In question 8 of Exercise 9a you will have established that

$$C^{-1}MC = D, \text{ where } C = \begin{pmatrix} 1 & 1 \\ 1 & 2 \end{pmatrix}, \quad M = \begin{pmatrix} 4 & -1 \\ 2 & 1 \end{pmatrix}, \quad D = \begin{pmatrix} 3 & 0 \\ 0 & 2 \end{pmatrix}.$$

Since $\qquad M^5 = CD^5C^{-1},$

$$M^5 = \begin{pmatrix} 1 & 1 \\ 1 & 2 \end{pmatrix} \begin{pmatrix} 3^5 & 0 \\ 0 & 2^5 \end{pmatrix} \begin{pmatrix} 2 & -1 \\ -1 & 1 \end{pmatrix}$$

$$= \begin{pmatrix} 2(3^5)-(2^5) & -(3^5)+(2^5) \\ 2(3^5)-2(2^5) & -(3^5)+2(2^5) \end{pmatrix}.$$

You probably noticed that for each of questions 8 and 9 of Exercise 9a, the matrix C had columns which were eigenvectors of the matrix M, and also that the product matrix $C^{-1}MC$ was a diagonal matrix whose diagonal entries were eigenvalues of M. We shall now prove that these examples illustrate a general result, namely that $MC = CD$.

THEOREM 9.2. *If a two by two matrix M has eigenvectors $\mathbf{e}_1$, $\mathbf{e}_2$ corresponding to eigenvalues k_1, k_2 respectively, then $MC = CD$, where C is the matrix $(\mathbf{e}_1\ \mathbf{e}_2)$, and $D = \begin{pmatrix} k_1 & 0 \\ 0 & k_2 \end{pmatrix}$.*

Proof. We need not specify the entries of M. By the rules of matrix multiplication, it is clear that

$$MC = (M\mathbf{e}_1, M\mathbf{e}_2).$$

Therefore $\qquad MC = (k_1\mathbf{e}_1, k_2\mathbf{e}_2)$, since $M\mathbf{e}_1 = k_1\mathbf{e}_1$, $M\mathbf{e}_2 = k_2\mathbf{e}_2$.

Now $\qquad CD = (\mathbf{e}_1, \mathbf{e}_2)\begin{pmatrix} k_1 & 0 \\ 0 & k_2 \end{pmatrix}.$

We observed in Chapter 8 that multiplying a matrix such as C on the right by an elementary matrix such as D results in elementary column operations on the columns of the matrix such as C. In this case, the columns of C will be multiplied by k_1, k_2 respectively.

That is,

$$MC = (k_1\mathbf{e}_1, k_2\mathbf{e}_2).$$

This establishes that

$$MC = CD.$$

The theorem implies that as long as C has an inverse, C^{-1}, then $C^{-1}MC = D$. Under what conditions will C have an inverse? In Chapter 6 we observed that if a matrix is singular its rows are linearly dependent, as are its columns. Conversely, if a matrix is non-singular its rows are linearly independent, as are its columns. For example, the matrix C of question 8 of Exercise 9a is non-singular and its columns are linearly independent; the same can be said of the matrix C of question 9.

We shall now prove that these examples illustrate a general result, namely that the eigenvectors corresponding to distinct (i.e. different) eigenvalues of a matrix are linearly independent. This implies that if a two by two matrix M has two distinct eigenvalues, then we may construct a non-singular matrix C whose columns are linearly independent eigenvectors of M.

THEOREM 9.3. *Eigenvectors corresponding to distinct eigenvalues of a two by two matrix are linearly independent.*

Proof. Let $\mathbf{e}_1$, $\mathbf{e}_2$ be eigenvectors corresponding to distinct eigenvalues k_1, k_2 of the matrix M. We shall show that the only values of a and b satisfying

$$a\mathbf{e}_1 + b\mathbf{e}_2 = \mathbf{0} \qquad \text{(A)}$$

are $a = b = 0$.

Multiplying the above equation by the matrix M gives

$$\begin{aligned} aM\mathbf{e}_1 + bM\mathbf{e}_2 &= M\mathbf{0} \\ &= \mathbf{0}. \end{aligned}$$

Therefore $ak_1\mathbf{e}_1 + bk_2\mathbf{e}_2 = \mathbf{0}$, since $M\mathbf{e}_1 = k_1\mathbf{e}_1$, $M\mathbf{e}_2 = k_2\mathbf{e}_2$.

Multiplying the original equation (A) by k_1 gives

$$ak_1\mathbf{e}_1 + bk_1\mathbf{e}_2 = \mathbf{0}.$$

Subtracting the two last equations, we have

$$b(k_1 - k_2)\mathbf{e}_2 = \mathbf{0}.$$

Since $\mathbf{e}_2 \neq \mathbf{0}$, and $k_1 - k_2 \neq 0$ because k_1 and k_2 are distinct, therefore $b = 0$.

Since $\mathbf{e}_1 \neq \mathbf{0}$, it follows that $a = 0$.

This shows that $\mathbf{e}_1$ and $\mathbf{e}_2$ are linearly independent.

Note. Theorems 9.2 and 9.3 can be generalised to apply to any square matrix.

The process of forming a diagonal matrix $C^{-1}MC$ is called diagonalisation of M.

EXAMPLE (viii). Diagonalise the matrix $A = \begin{pmatrix} 3 & 1 \\ 1 & 3 \end{pmatrix}$.

In answering question 2 of Exercise 9a you probably found that the eigenvalues of A are 4 and 2.

The theorems we have proved show that a matrix C may be found so that

$$C^{-1}AC = \begin{pmatrix} 4 & 0 \\ 0 & 2 \end{pmatrix}.$$

Check that if the matrix $\begin{pmatrix} 1 & -1 \\ 1 & 1 \end{pmatrix}$ is substituted for C, the above equality holds.

Symmetric Matrices

Notice that the eigenvectors corresponding to distinct eigenvalues of the matrix A of Example (viii) are orthogonal. In fact, the eigenvectors of any symmetric matrix corresponding to distinct eigenvalues are orthogonal, as we shall now prove.

THEOREM 9.4. *Eigenvectors corresponding to distinct eigenvalues of a symmetric matrix are orthogonal.*

Proof. Consider eigenvectors $\mathbf{e}_1$, $\mathbf{e}_2$ corresponding to distinct eigenvalues k_1, k_2 of a symmetric matrix A, so that

$$k_1\mathbf{e}_1 = A\mathbf{e}_1 \quad \text{and} \quad k_1\mathbf{e}_1' = \mathbf{e}_1'A'.$$

(See section **Transpose of a Matrix** in Chapter 6.) The inner product of $\mathbf{e}_1$ and $\mathbf{e}_2$ is the product $\mathbf{e}_1'\mathbf{e}_2$.

Now
$$\begin{aligned} k_1\mathbf{e}_1'\mathbf{e}_2 &= \mathbf{e}_1'A'\mathbf{e}_2 \\ &= \mathbf{e}_1'A\mathbf{e}_2, \quad \text{since} \quad A = A', \\ &= \mathbf{e}_1'k_2\mathbf{e}_2, \quad \text{since} \quad A\mathbf{e}_2 = k_2\mathbf{e}_2, \\ &= k_2\mathbf{e}_1'\mathbf{e}_2. \end{aligned}$$

Thus $(k_1-k_2)\mathbf{e}_1'\mathbf{e}_2 = 0$.
Since k_1, k_2 are distinct, $k_1-k_2 \neq 0$.
Therefore $\mathbf{e}_1'\mathbf{e}_2 = 0$.
That is, $\mathbf{e}_1$ and $\mathbf{e}_2$ are orthogonal.

Check that for all three symmetric matrices you investigated in Exercise 9a, the eigenvectors are mutually orthogonal. This means that the matrix C whose columns are eigenvectors of the symmetric matrix will be orthogonal, provided unit eigenvectors are selected for its columns.

Even if the eigenvalues of a three by three matrix are not distinct, it may still be possible to find three linearly independent eigenvectors and hence diagonalise the matrix. In fact, a symmetric matrix may always be diagonalised.

EXAMPLE (ix). Show that the matrix $M = \begin{pmatrix} -1 & 2 & 2 \\ 2 & 2 & -1 \\ 2 & -1 & 2 \end{pmatrix}$ has only two eigenvalues, but nevertheless three linearly independent eigenvectors. Hence diagonalise M.

The characteristic equation of M is

$$(-1-k)[(2-k)(2-k)-1]-2[2(2-k)+2]+-2[-2(2-k)] = 0.$$

This simplifies to $\quad (k-3)^2(k+3) = 0,$
yielding two eigenvalues, $k_1 = -3$, $k_2 = 3$.

If the vector $\mathbf{e}_1 = \begin{pmatrix} x_1 \\ y_1 \\ z_1 \end{pmatrix}$ satisfies $(M+3I)\mathbf{e}_1 = \mathbf{0}$, then three linear equations hold for x_1, y_1, z_1, whose augmented matrix is

$$\begin{pmatrix} 2 & 2 & 2 & 0 \\ 2 & 5 & -1 & 0 \\ 2 & -1 & 5 & 0 \end{pmatrix},$$

$$\sim \begin{pmatrix} 1 & 1 & 1 & 0 \\ 0 & 1 & -1 & 0 \\ 0 & 0 & 0 & 0 \end{pmatrix},$$

implying that z_1 is arbitrary, $y_1 = z_1$, and $x_1 = -2y_1$.

If the vector $\mathbf{e}_2 = \begin{pmatrix} x_2 \\ y_2 \\ z_2 \end{pmatrix}$ satisfies $(M-3I)\mathbf{e}_2 = \mathbf{0}$, then three linear equations hold for x_2, y_2, z_2, whose augmented matrix is

$$\begin{pmatrix} -4 & 2 & 2 & 0 \\ 2 & -1 & -1 & 0 \\ 2 & -1 & -1 & 0 \end{pmatrix},$$

$$\sim \begin{pmatrix} 1 & -\frac{1}{2} & -\frac{1}{2} & 0 \\ 0 & 0 & 0 & 0 \\ 0 & 0 & 0 & 0 \end{pmatrix},$$

implying that z_2 and y_2 are arbitrary, and $x_2 = \frac{1}{2}(y_2+z_2)$.

From this set of eigenvectors we may select two particular ones which are linearly independent—for instance, $\begin{pmatrix} 1 \\ 0 \\ 2 \end{pmatrix}$ and $\begin{pmatrix} 1 \\ 2 \\ 0 \end{pmatrix}$.

With these and the eigenvector $\begin{pmatrix} -2 \\ 1 \\ 1 \end{pmatrix}$ from the first set, we may construct the columns of a matrix $C = \begin{pmatrix} -2 & 1 & 1 \\ 1 & 0 & 2 \\ 1 & 2 & 0 \end{pmatrix}$.

Provided C has an inverse, we know by Theorem 9.2 generalised that

$$C^{-1}MC = \begin{pmatrix} -3 & 0 & 0 \\ 0 & 3 & 0 \\ 0 & 0 & 3 \end{pmatrix}.$$

Check that C is non-singular, so that the diagonalisation process may in fact be carried out. Notice that the three eigenvectors forming the columns of C are not all mutually perpendicular. Why does this not refute Theorem 9.4?

Summary of Chapter 9

We defined an eigenvector and an eigenvalue of a square matrix. We proved that if k is an eigenvalue of the square matrix M, then $\det(M-kI) = 0$. This equation is called the characteristic equation of the matrix M. We noted that an n by n matrix has up to n but not more than n eigenvalues.

We proved that a two by two matrix with two distinct eigenvalues may be diagonalised by means of a matrix C, whose columns are eigenvectors of M. This result may be generalised to apply to any square matrix.

We saw that eigenvectors corresponding to distinct eigenvalues of a symmetric matrix are orthogonal. We noted (without proof) that a symmetric matrix may always be diagonalised, even if its eigenvalues are not distinct.

Exercise 9b

1. Diagonalise the following matrices:

$$\text{(a)} \begin{pmatrix} 0{\cdot}3 & 0{\cdot}1 \\ 0{\cdot}7 & 0{\cdot}9 \end{pmatrix}, \quad \text{(b)} \begin{pmatrix} 3 & -6 & -4 \\ -6 & 4 & 2 \\ -4 & 2 & -1 \end{pmatrix}.$$

2. Calculate M^n, where M is the matrix of question 1(a), and n is a positive integer. Predict the limit of M^n as n becomes very large.

3. Why would you expect the sets of eigenvectors of the matrix of question 1(b) to be mutually perpendicular? Check that they are.

4. Show that the matrix $\begin{pmatrix} 2 & -1 & 1 \\ 3 & 3 & -2 \\ 4 & 1 & 0 \end{pmatrix}$ cannot be diagonalised.

5. Show that the vectors $\begin{pmatrix} 2 \\ 2 \\ -1 \end{pmatrix}$, $\begin{pmatrix} -1 \\ 2 \\ 2 \end{pmatrix}$ are both eigenvectors of the matrix $\begin{pmatrix} 4 & -2 & 4 \\ -2 & 1 & -2 \\ 4 & -2 & 4 \end{pmatrix}$. Show that although this matrix has only two eigenvalues, nevertheless three linearly independent eigenvectors may be found. Hence diagonalise the matrix.

6. If k is a real non-zero eigenvalue of the three by three matrix M, show that (a) k is an eigenvalue of the matrix M', and that (b) $1/k$ is an eigenvalue of the matrix M^{-1}. Deduce that the eigenvalues of a three by three orthogonal matrix are ± 1. Show how Example (ix) of this chapter can be used to illustrate this fact.

7. A four by four matrix M has eigenvectors $\mathbf{e}_1$, $\mathbf{e}_2$, $\mathbf{e}_3$, $\mathbf{e}_4$ corresponding to eigenvalues k_1, k_2, k_3, k_4. Show that $MC = CD$, where $C = (\mathbf{e}_1, \mathbf{e}_2, \mathbf{e}_3, \mathbf{e}_4)$, and D is an appropriate diagonal matrix.

8. Prove that eigenvectors corresponding to distinct eigenvalues of a three by three matrix are linearly independent.

CHAPTER 10

SOME APPLICATIONS OF EIGENVECTORS

Introduction

In this final chapter we shall outline some of the variety of problems which can be solved with recourse to eigenvectors. The calculation involved in solving each problem is relatively simple, but the theory behind the method of solution is sometimes complex. We shall, however, prove most of the theorems on which the theory depends.

Quadratic Forms

A quadratic form is a homogeneous quadratic function of several variables. For example, the expression

$$ax^2+by^2+2fxy,$$

where a, b and f are real numbers, is a quadratic form in x and y; and the expression

$$ax^2+by^2+cz^2+2fxy+2gxz+2hyz$$

is a quadratic form in the variables x, y and z.

Such quantities often occur in mathematics—for example, variance in statistics, the equation of a conic section or a quadric surface in coordinate geometry, moments of inertia in mechanics. Moreover, such a quantity may be expressed as a product involving a matrix and a vector. For example, if A is the matrix $\begin{pmatrix} a & f \\ f & b \end{pmatrix}$ and $\mathbf{v}$ the vector $\begin{pmatrix} x \\ y \end{pmatrix}$, then

$$ax^2+2fxy+by^2 = \mathbf{v}'A\mathbf{v}.$$

Also, if B is the matrix $\begin{pmatrix} a & f & g \\ f & b & h \\ g & h & c \end{pmatrix}$ and $\mathbf{w}$ the vector $\begin{pmatrix} x \\ y \\ z \end{pmatrix}$, then

$$ax^2+by^2+cz^2+2fxy+2gxz+2hyz = \mathbf{w}'B\mathbf{w}.$$

Notice that we selected A and B as symmetric matrices. Although other matrices M, N could have been selected to express the quadratic forms as $\mathbf{v}'M\mathbf{v}$, $\mathbf{w}'N\mathbf{w}$, there are advantages in selecting symmetric matrices, as will be seen later.

EXAMPLE (i). Express the quadratic forms (a) $2x^2-3xy+6y^2 = f(x, y)$, (b) $3x^2-12xy+4yz-8xz+4y^2-z^2 = g(x, y, z)$, (c) $ax^2+by^2+cz^2 = h(x, y, z)$, as products involving matrices and vectors.

(a) Let $\mathbf{v} = \begin{pmatrix} x \\ y \end{pmatrix}$. Then if $A = \begin{pmatrix} 2 & -\frac{3}{2} \\ -\frac{3}{2} & 6 \end{pmatrix}$, $f(x, y) = \mathbf{v}'A\mathbf{v}$.

Or, if $M = \begin{pmatrix} 2 & -1 \\ -2 & 6 \end{pmatrix}$, then $f(x, y) = \mathbf{v}'M\mathbf{v}$.

(b) Let $\mathbf{w} = \begin{pmatrix} x \\ y \\ z \end{pmatrix}$. Then if $B = \begin{pmatrix} 3 & -6 & -4 \\ -6 & 4 & 2 \\ -4 & 2 & -1 \end{pmatrix}$, $g(x, y, z) = \mathbf{w}'B\mathbf{w}$.

(c) Let $\mathbf{w} = \begin{pmatrix} x \\ y \\ z \end{pmatrix}$. Then if $D = \begin{pmatrix} a & 0 & 0 \\ 0 & b & 0 \\ 0 & 0 & c \end{pmatrix}$, $h(x, y, z) = \mathbf{w}'D\mathbf{w}$.

Diagonalisation

Notice that the matrix D of Example (i), part (c), is a diagonal matrix. The quadratic form $ax^2+by^2+cz^2$, having no terms in xy or yz or xz, is said to be a diagonal quadratic form.

It is sometimes useful to diagonalise a quadratic form—that is, to find an orthogonal linear mapping of vectors whose components are the variables that will map the form onto a diagonal form. In the following theorem we demonstrate how this may be accomplished. We shall see the point of expressing the form as $\mathbf{v}'A\mathbf{v}$, where A is a symmetric matrix.

THEOREM 10.1. *If A is a symmetric three by three matrix with distinct eigenvalues k_1, k_2, k_3 and corresponding unit eigenvectors $\mathbf{e}_1$, $\mathbf{e}_2$, $\mathbf{e}_3$, then the quadratic form $\mathbf{v}'A\mathbf{v}$, where $\mathbf{v} = \begin{pmatrix} x \\ y \\ z \end{pmatrix}$, is mapped by the mapping $\mathbf{v}_1 = U^{-1}\mathbf{v}$, where $U = (\mathbf{e}_1, \mathbf{e}_2, \mathbf{e}_3)$ and $\mathbf{v}_1 = \begin{pmatrix} x_1 \\ y_1 \\ z_1 \end{pmatrix}$, onto the diagonal form $k_1x_1^2+k_2y_1^2+k_3z_1^2$.*

Proof. By Theorem 9.4, U is an orthogonal matrix.

If $U^{-1}\mathbf{v} = \mathbf{v}_1$, then $\mathbf{v}=U\mathbf{v}_1$ (multiplying both sides on the left by U).

$$\text{Therefore } \mathbf{v}' = \mathbf{v}_1'U',$$
$$= \mathbf{v}_1'U^{-1}, \text{ because } U \text{ is orthogonal.}$$
$$\text{Therefore } \mathbf{v}'A\mathbf{v} = \mathbf{v}_1'U^{-1}AU\mathbf{v}_1.$$

$$\text{But, by Theorem 9.2 } U^{-1}AU = D = \begin{pmatrix} k_1 & 0 & 0 \\ 0 & k_2 & 0 \\ 0 & 0 & k_3 \end{pmatrix}.$$

$$\text{Therefore } \mathbf{v}'A\mathbf{v} = \mathbf{v}_1'D\mathbf{v}_1$$
$$= k_1x_1^2+k_2y_1^2+k_3z_1^2.$$

Note. This theorem can be generalised to apply to any symmetric n by n matrix and its associated quadratic form.

EXAMPLE (ii). What curve C is represented by the coordinate equation

$$2x^2-3xy+6y^2 = 1?$$

The equation may be written in the form

$$\mathbf{v}'A\mathbf{v} = 1,$$

where A and $\mathbf{v}$ are as defined in Example (i).

The eigenvalues of A satisfy the characteristic equation

$$(2-k)(6-k)-\tfrac{9}{4} = 0.$$

Thus
$$k^2-8k+\tfrac{39}{4} = 0,$$
which is satisfied by $k = \frac{13}{2}$ or $\frac{3}{2}$.

A unit eigenvector corresponding to $k_1 = \frac{3}{2}$ is $\mathbf{e}_1 = \dfrac{1}{\sqrt{10}}\begin{pmatrix}3\\1\end{pmatrix}$,

and one corresponding to $k_2 = \frac{13}{2}$ is $\mathbf{e}_2 = \dfrac{1}{\sqrt{10}}\begin{pmatrix}-1\\3\end{pmatrix}$.

By Theorem 10.1, we know that the mapping $\mathbf{v}_1 = U^{-1}\mathbf{v}$, where $U = \dfrac{1}{\sqrt{10}}\begin{pmatrix}3 & -1\\1 & 3\end{pmatrix}$ and $\mathbf{v}_1 = \begin{pmatrix}x_1\\y_1\end{pmatrix}$, maps the curve C onto the curve C_1 whose coordinate equation is

$$\tfrac{3}{2}x_1^2+\tfrac{13}{2}y_1^2 = 1.$$

This is recognisable as the equation of an ellipse whose major axis is $2\sqrt{\frac{2}{3}}$ and minor axis is $2\sqrt{\frac{2}{13}}$.

Since the matrix U is orthogonal, so is U^{-1}, which establishes that the mapping was an isometric one. Therefore the curve C is identical in size and shape to C_1.

Note. It was not necessary to calculate the eigenvectors $\mathbf{e}_1$, $\mathbf{e}_2$ in Example (ii). We confirmed thereby that U was an orthogonal matrix, but this was in fact implied by Theorem 10.1.

EXAMPLE (iii). Prove that the quadratic form $2x^2-3xy+6y^2$ never takes a negative value, whatever the real values of x and y. Find its minimum value subject to the condition $x^2+y^2 = 1$.

From Example (ii), we know that if

$$\sqrt{10}x_1 = 3x + y,$$
$$\sqrt{10}y_1 = x - 3y,$$

then $$2x^2 - 3xy + 6y^2 = \tfrac{3}{2}x_1^2 + \tfrac{13}{2}y_1^2.$$

Clearly, whatever the real values of x_1 and y_1, the expression $\frac{3}{2}x_1^2 + \frac{13}{2}y_1^2$ cannot be negative. Therefore neither can

$$2x^2 - 3xy + 6y^2.$$

The expression $x^2 + y^2$ represents the square of a length. Since the mapping performed is isometric, lengths are preserved, and therefore $x_1^2 + y_1^2 = x^2 + y^2 = 1$.

Subject to this condition, clearly the minimum value of $\frac{3}{2}x_1^2 + \frac{13}{2}y_1^2$ is obtained when $x_1 = 1$, $y_1 = 0$.

Thus the minimum value taken by the form is $\frac{3}{2}$.

Check that when $x_1 = 1$, $y_1 = 0$, $x = \dfrac{3}{\sqrt{10}}$, $y = \dfrac{1}{\sqrt{10}}$, and

$$2x^2 - 3xy + 6y^2 = \tfrac{3}{2}.$$

Note. A quadratic form which is positive for all the values of its variables, except when these are all zero, is called positive definite. Thus the form investigated in Examples (ii) and (iii) is positive definite.

EXAMPLE (iv). Find the shape of the surface S, whose equation is

$$3x^2 + 4y^2 - z^2 - 12xy - 8xz + 4yz = 1.$$

The equation may be written in the form

$$\mathbf{w}'B\mathbf{w} = 1,$$

where B and $\mathbf{w}$ are as defined in Example (i).

The eigenvalues of B satisfy the characteristic equation

$$(3-k)[(4-k)(-1-k)-4]+6[6(1+k)+8]-4[-12+4(4-k)] = 0.$$

Thus $$-k^3+6k^2+51k+44 = 0,$$
which is satisfied by $k = -1, -4$, or 11.

By Theorem 10.1, we know that there is an orthogonal matrix mapping the surface S onto the surface S_1, whose equation is

$$-x_1^2-4y_1^2+11z_1^2 = 1.$$

Reference to the section *The Central Quadrics* of Chapter 4 reminds us that this is the equation of a hyperboloid of two sheets.

Since the mapping is isometric, S is congruent to S_1.

EXAMPLE (v). Find the maximum and minimum values of the quadratic form $Q = 3x^2+4y^2-z^2-12xy-8xz+4yz$, subject to the condition $x^2+y^2+z^2 \leqslant 1$.

From Example (iv) we know that there is an orthogonal matrix mapping $\begin{pmatrix} x \\ y \\ z \end{pmatrix}$ onto $\begin{pmatrix} x_1 \\ y_1 \\ z_1 \end{pmatrix}$, so that

$$Q = -x_1^2-4y_1^2+11z_1^2.$$

Since this mapping preserves lengths, $x_1^2+y_1^2+z_1^2 = x^2+y^2+z^2 \leqslant 1$, and subject to this condition, the maximum value of Q is obtained when $x_1 = y_1 = 0$, $z_1 = 1$, and the minimum value when $x_1 = z_1 = 0$, $y_1 = 1$.

Therefore Q has a maximum value of 11 and a minimum value of -4 subject to the imposed conditions.

Exercise 10a

1. Find the shape of the curve whose equation is

$$7x^2+8xy+y^2 = 1.$$

2. Find the maximum and minimum values of the quadratic form

$$7x^2+8xy+y^2$$

subject to the condition $x^2+y^2 \leq 3$.

3. Show that the quadratic form

$$Q = 20x^2+9y^2+20z^2+32xz$$

is positive definite. Show that its maximum value is 36, subject to the condition $x^2+y^2+z^2 \leq 1$.

4. Find the shape of the surface whose equation is

$$20x^2+9y^2+20z^2+32xz = 36.$$

Recurring Processes

The problems we consider in this section each concern a chain of recurring events, for which the outcome of any particular event depends only on the outcome of the previous event. For instance, consider the rabbit population introduced in Example (iv) of Chapter 7. Here the events are the births, maturations and deaths, summarised in the transition matrix M; the outcomes are the annual rabbit populations, the vector $\mathbf{v}_n$ representing the population after n years. We showed in the example how $\mathbf{v}_n$ depends on the vector $\mathbf{v}_{n-1}$; in fact, $\mathbf{v}_n = M\mathbf{v}_{n-1}$.

Another example will concern a chemical experiment, where liquid and gas are present together in a container. Here the events will be the condensation of some of the gas to liquid, and the simultaneous evaporation of some liquid to gas; the outcomes will be the amounts of liquid and gas present at certain intervals of time.

EXAMPLE (vi). x_0 gm. of liquid and y_0 gm. of gas are together in an enclosed container. Every minute, $\frac{7}{10}$ of the liquid present vaporises to gas, and $\frac{1}{10}$ of the gas condenses to liquid. Predict the eventual proportions of liquid and gas to be expected.

It might be supposed that no steady state will be reached in which these proportions remain fixed, or that, if a steady state is reached, these proportions would depend on the initial proportions of liquid to gas, $x_0 : y_0$. However, our investigation of this recurring process will reveal that such suppositions are wrong.

Suppose that after n minutes x_n gm. of liquid and y_n gm. of gas are present in the container.

Let $$\mathbf{v}_n = \begin{pmatrix} x_n \\ y_n \end{pmatrix}.$$

Then
$$x_1 = 0{\cdot}3x_0 + 0{\cdot}1y_0,$$
$$y_1 = 0{\cdot}7x_0 + 0{\cdot}9y_0.$$

Or, we may write
$$\mathbf{v}_1 = M\mathbf{v}_0, \text{ where } M = \begin{pmatrix} 0{\cdot}3 & 0{\cdot}1 \\ 0{\cdot}7 & 0{\cdot}9 \end{pmatrix}.$$

In fact,
$$\begin{aligned} \mathbf{v}_n &= M\mathbf{v}_{n-1}, \\ &= M(M\mathbf{v}_{n-2}), \\ &= M^2\mathbf{v}_{n-2}, \end{aligned}$$

and, by extending the argument,
$$\mathbf{v}_n = M^n\mathbf{v}_0.$$

This result would seem to confirm the supposition that the eventual state depends on the initial state that determines the vector $\mathbf{v}_0$. But now let us examine M^n. This you have already done, if you have answered question 2 of Exercise 9b, using the relation $M = CDC^{-1}$ to obtain
$$\begin{aligned} M^n &= CD^nC^{-1} \\ &= \tfrac{1}{8}\begin{pmatrix} 1+7(0{\cdot}2)^n & 1-(0{\cdot}2)^n \\ 7-7(0{\cdot}2)^n & 7+(0{\cdot}2)^n \end{pmatrix}, \end{aligned}$$

and deducing that as n becomes very large, $(0{\cdot}2)^n$ becomes very small, so that M^n approaches $\tfrac{1}{8}\begin{pmatrix} 1 & 1 \\ 7 & 7 \end{pmatrix}$.

Now $\mathbf{v}_n = M^n\mathbf{v}_0$, so that we deduce that $\mathbf{v}_n$ approaches
$$\tfrac{1}{8}\begin{pmatrix} 1 & 1 \\ 7 & 7 \end{pmatrix}\begin{pmatrix} x_0 \\ y_0 \end{pmatrix} = \begin{pmatrix} \tfrac{1}{8}(x_0+y_0) \\ \tfrac{7}{8}(x_0+y_0) \end{pmatrix}.$$

That is, the proportions of liquid and gas present approach $\tfrac{1}{8}$ and $\tfrac{7}{8}$ respectively. Once these proportions are reached, they are maintained, for
$$\begin{pmatrix} 0{\cdot}3 & 0{\cdot}1 \\ 0{\cdot}7 & 0{\cdot}9 \end{pmatrix}\begin{pmatrix} \tfrac{1}{8} \\ \tfrac{7}{8} \end{pmatrix} = \begin{pmatrix} \tfrac{1}{8} \\ \tfrac{7}{8} \end{pmatrix}.$$

(This last equation is a consequence of the fact that the vector $\begin{pmatrix} \frac{1}{8} \\ \frac{7}{8} \end{pmatrix}$ is an eigenvector of the matrix M.)

In Example (vi) we have shown that, whatever the initial state, the same eventual steady state is reached. Stated algebraically, whatever the value of $\mathbf{v}_0$, $M^n\mathbf{v}_0$ approaches a certain eigenvector of M. This is generally true, as we now prove in Theorem 10.2.

THEOREM 10.2. *Suppose the two by two matrix M has 1 as the numerically larger of its eigenvalues, and* $\mathbf{v}$ *is any two-dimensional vector which is not an eigenvector of M. Then, as n becomes very large,* $M^n\mathbf{v}$ *approaches an eigenvector of M corresponding to the eigenvalue 1.*

Proof. Let $\mathbf{e}_1$, $\mathbf{e}_2$ be eigenvectors of M corresponding to the eigenvalues 1, k_2, respectively.

By Theorem 9.3, $\mathbf{e}_1$ and $\mathbf{e}_2$ are linearly independent. Therefore, as we proved in Example (xi) of Chapter 8, the vector $\mathbf{v}$ may be expressed in the form

$$\mathbf{v} = a\mathbf{e}_1 + b\mathbf{e}_2,$$

where a and b are real and neither is zero, since $\mathbf{v}$ is not an eigenvector.

Thus

$$\begin{aligned} M^n\mathbf{v} &= aM^n\mathbf{e}_1 + bM^n\mathbf{e}_2 \\ &= a\mathbf{e}_1 + bk_2^n\mathbf{e}_2, \quad \text{since} \quad M\mathbf{e}_1 = \mathbf{e}_1 \quad \text{and} \quad M\mathbf{e}_2 = k_2\mathbf{e}_2. \end{aligned}$$

Since $|k_2| < 1$, as n becomes very large k_2^n approaches zero. Therefore $M^n\mathbf{v}$ approaches $a\mathbf{e}_1$, an eigenvector corresponding to the eigenvalue of 1.

EXAMPLE (vii). For the rabbit population introduced in Example (iv) of Chapter 7, show that whatever the initial population is, the population will approach a state in which there are three times as many infants as adults, with a population increase of $33\frac{1}{3}\%$ per year.

In Example (iv) of Chapter 7 we showed that the eigenvectors of the transition matrix $M = \begin{pmatrix} 0 & 4 \\ \frac{1}{3} & \frac{1}{3} \end{pmatrix}$ are $a\mathbf{e}_1$ and $b\mathbf{e}_2$, for all real non-zero a and b, where $\mathbf{e}_1 = \begin{pmatrix} 3 \\ 1 \end{pmatrix}$, $\mathbf{e}_2 = \begin{pmatrix} 4 \\ -1 \end{pmatrix}$ corresponding to the eigenvalues $\frac{4}{3}$ and -1.

Clearly, an initial population cannot be represented by a vector from the set $b\mathbf{e}_2$ which contains a negative entry; and if it is represented by a vector from the set $a\mathbf{e}_1$, then, as we have already shown in the same example, the age distribution can be expected to remain in the same ratio and the population to increase by $33\frac{1}{3}\%$ every year.

If, however, the initial population is represented by $\mathbf{v}$, where

$$\mathbf{v} = a\mathbf{e}_1 + b\mathbf{e}_2, \quad a \text{ and } b \text{ real and non-zero,}$$

then we need to investigate $M^n\mathbf{v}$ as n becomes large.

The matrix $\frac{3}{4}M$ has the same eigenvectors as M, and eigenvalues $1, -\frac{3}{4}$.

Therefore, by Theorem 10.2, $(\frac{3}{4}M)^n\mathbf{v}$ approaches the vector $\begin{pmatrix} 3a \\ a \end{pmatrix}$, for some real a, as n becomes large.

We deduce that the ratio of the entries of the vector $M^n\mathbf{v}$ approaches $3:1$, although the actual entries of this vector will become larger and larger as n becomes larger.

We have already shown that for an age distribution in this ratio $3:1$, the population can be expected to increase by $33\frac{1}{3}\%$ per year.

Probability

The concept of events and their outcomes is basic to the theory of probability. Provided that the possible outcomes of an event may be counted and considered as equally likely, we define the probability of a certain result to an event to be the number of outcomes which produce this result divided by the total number of possible outcomes. For example, we may consider throwing an unbiased die as an event with six equally possible outcomes. The probability of obtaining a six is then $\frac{1}{6}$, and the probability of obtaining a score higher than four is $\frac{2}{6}$, or $\frac{1}{3}$.

Probability may also be deduced from statistics. Suppose a survey reveals that out of 1000 births 504 are boys. It may then be useful to assume that the probability of a baby being born a boy is 0·504. Or, suppose that John is observed to stay at home on average one evening out of three. We might say that the probability of his staying at home any one evening is $\frac{1}{3}$. You will observe that, however the probability of a certain result is obtained, it will be a number lying between 0 and 1. A value of 0 would signify that the result was impossible to obtain (for instance, throwing a seven with a die), and a value of 1 that the result is certain to be obtained (for instance, scoring less than seven with a die).

The sum of the probabilities of all possible outcomes of an event must be 1. For example, with the same survey data, the probability of a baby being born a girl must be taken as 0·496; and the probability of John going out on any one evening must be $\frac{2}{3}$. In fact, to ascertain the probability of one result or another we may add the probabilities of each result. The probability of John either staying at home or going out is $\frac{1}{3}+\frac{2}{3}$, or 1, signifying he is certain to do one of these. The probability of throwing either a five or a six with a die is $\frac{1}{6}+\frac{1}{6}$, or $\frac{1}{3}$. In general, if the probabilities of two mutually exclusive results are p and q, then the probability of one result or the other is $p+q$.

Now consider two events—for example, throwing two dice. The possible outcomes of the two throws can be listed: 1 followed by 1, 1 followed by 2, 1 followed by 3, etc., giving 36 possible outcomes. The probability of throwing a six with each die is thus $\frac{1}{36}$, which is $\frac{1}{6}\times\frac{1}{6}$. What is the probability of scoring more than four with each die? The outcomes which produce this result are: 5 followed by 5, 5 followed by 6, 6 followed by 5, 6 followed by 6, giving 4 outcomes in all. Therefore the required probability is $\frac{4}{36}$, which is $\frac{2}{6}\times\frac{2}{6}$. These examples illustrate that if the probabilities of two results of different events are p and q, then the probability of both results occurring is pq.

The problems we are about to consider concern chains of events for which the probability of a particular outcome to an event depends on the outcome of the previous event. For instance, the probability of John's going out on a particular evening might well depend on whether he went out the previous evening.

EXAMPLE (viii). If John stays at home one evening, the probability that he stays the next evening is $\frac{1}{3}$. If he goes out, the probability that he stays home the next evening is $\frac{3}{4}$. If the probability of his staying at home on the evening of January 1 is p_0, predict the proportion of evenings of the year that he is likely to spend at home.

Let $\mathbf{v}_n = \begin{pmatrix} p_n \\ q_n \end{pmatrix}$, where p_n is the probability of John's staying at home on the nth evening after January 1, and $p_n+q_n = 1$.

If John stays home on January 2, then either he stayed home on January 1 and 2, for which the probability is $\frac{1}{3}p_0$, or he went out on January 1 and stays home on January 2, for which the probability is $\frac{3}{4}q_0$. Therefore p_1, the probability that he stays home on January 2, is given by

$$p_1 = \tfrac{1}{3}p_0+\tfrac{3}{4}q_0.$$

Similarly, q_1, the probability that he goes out on January 2, is given by

$$q_1 = \tfrac{2}{3}p_0+\tfrac{1}{4}q_0.$$

That is, $\mathbf{v}_1 = M\mathbf{v}_0$, where $M = \begin{pmatrix} \frac{1}{3} & \frac{3}{4} \\ \frac{2}{3} & \frac{1}{4} \end{pmatrix}$,

and
$$\begin{aligned} \mathbf{v}_n &= M\mathbf{v}_{n-1} \\ &= M^n\mathbf{v}_0. \end{aligned}$$

The characteristic equation of M is

$$(\tfrac{1}{3}-k)(\tfrac{1}{4}-k)-\tfrac{1}{2} = 0,$$

which is satisfied by $k = 1$ or $k = -\frac{5}{12}$.

Since M has 1 as the numerically larger of its eigenvalues, by Theorem 10.2 we know that as n becomes large, $M^n\mathbf{v}_0$ approaches an eigenvector corresponding to the eigenvalue of 1.

The vector $\mathbf{e} = \begin{pmatrix} x \\ y \end{pmatrix}$ is such an eigenvector, satisfying $(M-I)\mathbf{e} = \mathbf{0}$, if

$$-\tfrac{2}{3}x+\tfrac{3}{4}y = 0.$$

Thus $\mathbf{v}_n = M^n\mathbf{v}_0$ approaches a vector of the form $\begin{pmatrix} 9a \\ 8a \end{pmatrix}$ (where a is in fact $\frac{1}{17}$, since the entries of every $\mathbf{v}_n$ add to 1).

From this we predict that John may be expected eventually to spend $\frac{9}{17}$ of his evenings at home, and $\frac{8}{17}$ of them going out.

Check that $\mathbf{v}_0$ could not be an eigenvector corresponding to the eigenvalue $-\frac{5}{12}$ by showing that such eigenvectors have a negative entry.

Genetics

In the theory of genetics, the chain of events under consideration is the births of successive generations of animals, and the outcomes are the exhibition of certain inherited traits. Clearly here the outcome of one event depends on the outcome of the previous event (the birth of the parent). The theory is that genes governing certain individual traits, such as colour of eyes, are possessed in pairs by every animal. One of this pair is inherited from each parent. Genes may be of the dominant type, which we shall denote by G, or recessive, denoted by g. Animals possessing the gene combination GG or Gg will exhibit the same dominant trait, which might be brown eyes, while those possessing the gg combination will exhibit the recessive one, which might be blue eyes. Under this system, even if two parents exhibit the dominant trait, they might in fact both possess the recessive gene and produce an offspring exhibiting the recessive trait. But if both parents exhibit the recessive trait, they cannot produce an offspring exhibiting the dominant trait. In Fig. 10.1 we illustrate the four equally likely possible genetic results of a mating of two hybrids (those possessing the Gg combination), and also the possible results of a mating between a dominant and a hybrid.

EXAMPLE (ix). Hybrid type males are mated with x_0 dominant females, y_0 hybrid females and z_0 recessive females, and each female is expected to produce one offspring. Calculate x_1, y_1 and z_1, the expected numbers of dominant, hybrid and recessive offspring from the matings. If this new generation and successive generations con-

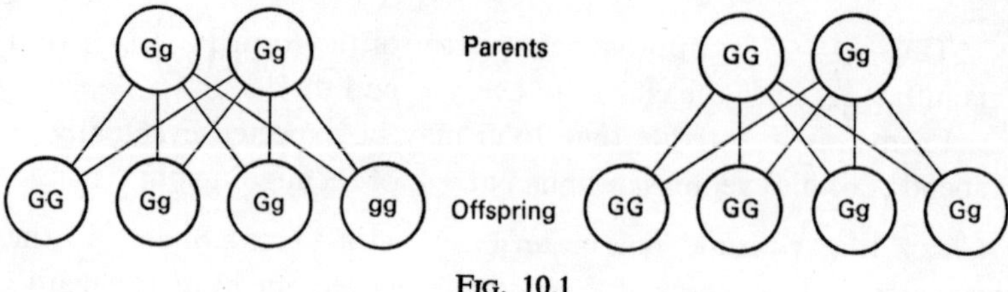

FIG. 10.1

tinue to mate with hybrid types, predict the eventual distribution of types after many generations.

We see from our illustration that for hybrid–dominant matings, $\frac{1}{2}$ of the offspring may be expected to be hybrid and $\frac{1}{2}$ dominant; and that for hybrid–hybrid matings $\frac{1}{4}$ of the offspring may be expected to be dominant, $\frac{1}{4}$ recessive and $\frac{1}{2}$ hybrid. Clearly, for hybrid–recessive matings, $\frac{1}{2}$ the offspring may be expected to be hybrid and $\frac{1}{2}$ recessive.

We deduce:

$$\begin{aligned} x_1 &= \tfrac{1}{2}x_0+\tfrac{1}{4}y_0, \\ y_1 &= \tfrac{1}{2}x_0+\tfrac{1}{2}y_0+\tfrac{1}{2}z_0, \\ z_1 &= \qquad\;\; \tfrac{1}{4}y_0+\tfrac{1}{2}z_0. \end{aligned}$$

Let $\mathbf{v}_n = \begin{pmatrix} x_n \\ y_n \\ z_n \end{pmatrix}$, where x_n, y_n, z_n denote the numbers of the various types in the nth generation.

Then $\mathbf{v}_1 = M\mathbf{v}_0$, where $M = \begin{pmatrix} \frac{1}{2} & \frac{1}{4} & 0 \\ \frac{1}{2} & \frac{1}{2} & \frac{1}{2} \\ 0 & \frac{1}{4} & \frac{1}{2} \end{pmatrix}$,

and

$$\begin{aligned} \mathbf{v}_n &= M\mathbf{v}_{n-1} \\ &= M^n\mathbf{v}_0. \end{aligned}$$

The characteristic equation of M is

$$(\tfrac{1}{2}-k)[(\tfrac{1}{2}-k)(\tfrac{1}{2}-k)-\tfrac{1}{8}]-\tfrac{1}{4}[\tfrac{1}{2}(\tfrac{1}{2}-k)] = 0,$$

that is, $$(\tfrac{1}{2}-k)(k^2-k) = 0,$$

which is satisfied by $$k = 1, \tfrac{1}{2} \text{ or } 0.$$

Check that corresponding to $k_1 = 1$ there is a set of eigenvectors parallel to $\mathbf{e}_1 = \begin{pmatrix} 1 \\ 2 \\ 1 \end{pmatrix}$, corresponding to $k_2 = \frac{1}{2}$ there is a set parallel to $\mathbf{e}_2 = \begin{pmatrix} 1 \\ 0 \\ -1 \end{pmatrix}$, and corresponding to $k_3 = 0$ there is a set parallel to $\mathbf{e}_3 = \begin{pmatrix} 1 \\ -2 \\ 1 \end{pmatrix}$.

In answering question 7 of Exercise 8b, you showed that if $\mathbf{e}_1$, $\mathbf{e}_2$, $\mathbf{e}_3$ are any linearly independent three-dimensional vectors, then any other three-dimensional vector $\mathbf{v}_0$ may be written in the form

$$\mathbf{v}_0 = a\mathbf{e}_1 + b\mathbf{e}_2 + c\mathbf{e}_3.$$

Now the components of $\mathbf{v}_0$ of this example are all positive. Therefore $a \neq 0$, for any vector of the form $b\mathbf{e}_2 + c\mathbf{e}_3$ will contain at least one negative component.

Therefore $M^n\mathbf{v}_0 = aM^n\mathbf{e}_1 + bM^n\mathbf{e}_2 + cM^n\mathbf{e}_3$

$$= a\mathbf{e}_1 + (\tfrac{1}{2})^n\mathbf{e}_2 + \mathbf{0}, \quad \text{since} \quad M\mathbf{e}_2 = \tfrac{1}{2}\mathbf{e}_2, \text{ etc.}$$

As n becomes large, $M^n\mathbf{v}_0$ clearly approaches $a\mathbf{e}_1$, indicating that we may expect the eventual distribution of types to be in the proportions $1 : 2 : 1$.

Summary of Chapter 10

We have investigated quadratic forms, and seen how these may be diagonalised, using the theory of eigenvectors we established in Chapter 9. We were enabled to find the shape of some curves and surfaces represented by second-degree coordinate equations, and to evaluate some maximum and minimum values of quadratic forms.

We investigated some recurring processes which could be summarised by a vector equation of the type $\mathbf{v}_n = M^n\mathbf{v}_0$, and we saw that for certain matrices M the vector $\mathbf{v}_n$ could be predicted to approach a certain eigenvector of M as n becomes large.

We introduced the theory of probability, and applied our matrix techniques to some probability recurrence situations and to a problem in genetics.

Outlook

We based our initial ideas of vectors on geometric concepts. From Chapter 7 onwards we applied matrix techniques in non-geometric contexts, and we replaced our geometric definitions by algebraic ones. However, geometric interpretations could still be relevant.

In the last two chapters we have used such terms as *orthogonal*, *parallel* and *length* in connection with two- and three-dimensional vectors which do not necessarily have any geometric relevance. Similarly, direction, orthogonality and length can be given a plausible interpretation for four-dimensional and n-dimensional vectors.

We have confined ourselves in this book to a discussion of vectors and matrices whose entries are real numbers. The ideas we have discussed are easily generalised to vectors and matrices with complex entries. If you are interested in developments such as are hinted at here, you will enjoy reading some of the books listed in the Bibliography on p. 174.

Exercise 10b

1. In an experiment on heredity in mice, it is found that $\frac{9}{10}$ of the offspring of white mice are white and $\frac{1}{10}$ are piebald, and that $\frac{1}{4}$ of the offspring of piebald mice are white and $\frac{3}{4}$ are piebald. Assuming that the proportions of white and piebald mice in the parent population are x_0, y_0, derive a relation between the vectors $\mathbf{v}_0 = \begin{pmatrix} x_0 \\ y_0 \end{pmatrix}$ and $\mathbf{v}_n = \begin{pmatrix} x_n \\ y_n \end{pmatrix}$, where x_n, y_n are the proportions of white and piebald mice that may be expected in the nth generation. Deduce the eventual proportion of the population expected to be piebald.

2. For the matrix $M = \begin{pmatrix} p & q \\ 1-p & 1-q \end{pmatrix}$, where $0 < p < 1$, $0 < q < 1$, show that the eigenvalues of M are 1 and $p-q$. Hence show that for any vector $\mathbf{v}$ not parallel to $\begin{pmatrix} 1 \\ -1 \end{pmatrix}$, as n becomes large the vector $M^n\mathbf{v}$ approaches some vector parallel to $\begin{pmatrix} q \\ 1-p \end{pmatrix}$.

3. In a study on population migration, it is found that 2% of the population move annually from rural to urban areas, while 1% move from urban to rural areas. Predict the eventual population distribution between rural and urban areas if this trend continues. (The result of question 2 could curtail your working.)

4. A rat is learning to find food in a maze. If he finds the right way to the food, there is a probability of $\frac{3}{4}$ that he will remember that way correctly for his next journey. But if he does not find the right way, then there is a probability of only $\frac{1}{6}$ that he will find the right way next time. Calculate the proportion of his attempts for which you would expect him to find the right way to the food.

5. A hybrid and a dominant type animal exhibit the same traits. Devise a means of testing whether an animal showing these traits is in fact dominant or hybrid. (*Hint:* devise a breeding experiment.)

Predict the eventual expected distribution of types after continually mating all types for many generations with recessive types.

6. In Example (vii) of this chapter we used the transition matrix $\begin{pmatrix} 0 & 4 \\ \frac{1}{3} & \frac{1}{3} \end{pmatrix}$ for rabbit population and showed that such a population could eventually be expected to increase by $33\frac{1}{3}\%$ per year. Suppose now a hunting policy is devised which is expected to destroy $\frac{2}{5}$ of the adult population at the beginning of every year. Show that the rabbit population can then be expected to remain static. (*Hint:* construct a new transition matrix to fit the altered conditions and consider its eigenvalues.)

BIBLIOGRAPHY

More about matrices

BIRKHOFF, G. and MACLANE, S., *A Survey of Modern Algebra*, Macmillan, 1965.
COHN, P. M., *Linear Equations*, Routledge & Kegan Paul, 1958.
FINKBEINER, D. T., *Introduction to Matrices and Linear Transformations*, Freeman, 1960.
HALMOS, P. R., *Finite Dimensional Vector Spaces*, Van Nostrand, 1958.
MOORE, J. T., *Elements of Linear Algebra and Matrix Theory*, McGraw-Hill, 1968.
SAWYER, W. W., *A Concrete Approach to Abstract Algebra*, Freeman, 1959.

More about vectors

MACBEATH, A. M., *Elementary Vector Algebra*, O.U.P., 1964.

More about probability and other applications

FELLER, W., *Introduction to Probability Theory and its Applications*, Wiley, 1967.
KEMENY, MIRKIL, SNELL and THOMPSON, *Finite Mathematical Structures*, Prentice-Hall, 1964.
KEMENY and SNELL, *Mathematical Models in the Social Sciences*, Blaisdell, 1962.

Suitable for children

JOHNSON, D. and GLENN, W., *Basic Concepts of Vectors* (Exploring Mathematics on Your Own, No. 16), Murray, 1964.
MATTHEWS, G., *Matrices* (Contemporary School Mathematics Series), Arnold, 1964.

ANSWERS TO THE EXERCISES

Exercise 1a

2. $(1\cdot4)^2 = 1\cdot96$, $(1\cdot5)^2 = 2\cdot25$, so $1\cdot4$ is nearer than $1\cdot5$ to $\sqrt{2}$. $(1\cdot41)^2 = 1\cdot9881$, $(1\cdot42)^2 = 2\cdot0164$, so both $1\cdot41$ and $1\cdot42$ are within $0\cdot01$ of $\sqrt{2}$.

3.	*Integer solutions*	*Real number solutions*
(a)	-3	-3
(b)	Every integer	Every real number
(c)	0	0
(d)	None	$\frac{1}{3}$
(e)	$2, -2$	$2, -2$
(f)	None	$\frac{1}{4}, -\frac{1}{4}$.

4. All of them.

Exercise 1b

1. (a), (c) and (e) are meaningless.

2. $\mathbf{v} = \begin{pmatrix} 1 \\ -3 \end{pmatrix}$, $\mathbf{w} = \begin{pmatrix} 2 \\ -5 \end{pmatrix}$, $\mathbf{u} = \mathbf{w}$.

3. $\overline{PQ} = \begin{pmatrix} -4 \\ 2 \end{pmatrix}$, $\overline{QR} = \begin{pmatrix} -2 \\ -4 \end{pmatrix}$, $\overline{RP} = \begin{pmatrix} 6 \\ 2 \end{pmatrix}$.

Notice that $\overline{PQ} = 2\overline{OB}$, $\overline{RP} = 2\overline{OA}$, $\overline{QR} = 2\overline{AB}$, $\overline{PQ}+\overline{QR}+\overline{RP} = \mathbf{0}$.

4. Parallelogram. $\overline{CD} = -\mathbf{w}$, $\overline{BD} = \mathbf{v}-\mathbf{w}$, $\overline{CA} = -\mathbf{v}-\mathbf{w}$.

5. Trapezium. $\overline{CD} = \mathbf{v}-\mathbf{w}$, $\overline{BD} = 3\mathbf{v}-\mathbf{w}$, $\overline{CA} = -2\mathbf{v}-\mathbf{w}$.

6. 130 m.p.h., bearing θ west of north, where $\tan\theta = \frac{5}{12}$.

8. $\overline{AM} = \frac{1}{2}(\mathbf{b}+\mathbf{c})$, $\overline{AP} = \frac{2}{3}\mathbf{c}+\frac{1}{3}\mathbf{b}$, $\overline{AQ} = \frac{1}{3}\mathbf{c}+\frac{2}{3}\mathbf{b}$.

Exercise 2a

1. (a), (b) and (c) are meaningless.

3. $\mathbf{c}\cdot\mathbf{c} = \mathbf{a}^2+\mathbf{b}^2+2\mathbf{a}\cdot\mathbf{b}$, $\mathbf{d}\cdot\mathbf{d} = \mathbf{a}^2+\mathbf{b}^2-2\mathbf{a}\cdot\mathbf{b}$.

4. (a) $\frac{1}{5}\begin{pmatrix} 4 \\ 3 \end{pmatrix}$, (b) $\begin{pmatrix} 1 \\ 0 \end{pmatrix}$, (c) $\frac{1}{\sqrt{10}}\begin{pmatrix} 3 \\ 1 \end{pmatrix}$.

5. $\overline{OM}\cdot\overline{PQ} = 0$. Triangle OPQ is isosceles.

6. $\frac{5}{2}$. Triangle AOB has base $AB = 5$, and height 1.

Exercise 2b

4. Let $\mathbf{a} = \begin{pmatrix}2\\3\end{pmatrix}$. Then (a) $\mathbf{v} = \mathbf{a}+r\mathbf{s}$, $x = 2$;

(b) $\mathbf{v} = \mathbf{a}+r\mathbf{t}$, $2x+3y = 13$;

(c) $\mathbf{v} = \mathbf{a}+r\mathbf{b}$, where $\mathbf{b} = \begin{pmatrix}2\\2\end{pmatrix}$, $x = y-1$.

5. Angle between (a) and (b) is α, where $\cos\alpha = \dfrac{2}{\sqrt{13}}$.

Angle between (a) and (c) is 45°.

Angle between (b) and (c) is β, where $\cos\beta = \dfrac{1}{\sqrt{26}}$.

6. $\sqrt{5}$; $2\sqrt{5}$.

Exercise 3a

2. $\overline{OP} = \begin{matrix}-3\\2,\\1\end{matrix}$ $OP = 14$, $\cos POQ = \dfrac{-3}{\sqrt{14}}$, triangle $OPQ = 1$.

3. $AC = \sqrt{20}$, $AB = \sqrt{6}$, $\cos BAC = \dfrac{3}{\sqrt{30}}$, triangle $ABC = \sqrt{21}$.

6. Not collinear.

7. Points are coplanar. $OABC$ is a rectangle.

Exercise 3b

1. $OA: x = 0, \quad \dfrac{y-1}{1} = \dfrac{z-5}{5}. \quad OB: \dfrac{x-1}{1} = \dfrac{y-3}{3} = \dfrac{z-4}{4}.$

$AC: \dfrac{x}{1} = \dfrac{y-1}{3} = \dfrac{z-5}{4}. \quad BC: x = 1, \dfrac{y-3}{1} = \dfrac{z-4}{5}.$

C is (1, 4, 9). OC, AB are perpendicular. $OABC$ is a rhombus.

4. $3y+4z = 25$.

5. $\begin{pmatrix}1\\1\\2\end{pmatrix}, \begin{pmatrix}2\\0\\2\end{pmatrix}; \quad \dfrac{2}{\sqrt{6}}, \dfrac{5}{2\sqrt{2}}.$

6. $\dfrac{x}{1} = \dfrac{y+3}{1} = \dfrac{z-\frac{5}{2}}{-1}$; 30°.

Exercise 4a

1. (a) One–one, topological; (b) one–one, isometric; (c) many–one; (d) one–one, affine.

2. (a) $\alpha^{-1} = \alpha$; (b) β has no inverse; (c) $\gamma^{-1} = \gamma$.

4. (a) Translation through twice the distance between the planes, in a direction perpendicular to them.

(b) Rotation through twice the angle between the planes.

Exercise 4b

1.

	Image of **i**	*Image of* **j**	*Image of* $\mathbf{v} = \begin{pmatrix} x \\ y \end{pmatrix}$	*Matrix*
(a)	$\begin{pmatrix} 1 \\ 0 \end{pmatrix}$	$\begin{pmatrix} 0 \\ -1 \end{pmatrix}$	$\begin{pmatrix} x \\ -y \end{pmatrix}$	$\begin{pmatrix} 1 & 0 \\ 0 & -1 \end{pmatrix}$,
(b)	$\begin{pmatrix} 0 \\ 1 \end{pmatrix}$	$\begin{pmatrix} 1 \\ 0 \end{pmatrix}$	$\begin{pmatrix} y \\ x \end{pmatrix}$	$\begin{pmatrix} 0 & 1 \\ 1 & 0 \end{pmatrix}$,
(c)	$\begin{pmatrix} -1 \\ 0 \end{pmatrix}$	$\begin{pmatrix} 0 \\ -1 \end{pmatrix}$	$\begin{pmatrix} -x \\ -y \end{pmatrix}$	$\begin{pmatrix} -1 & 0 \\ 0 & -1 \end{pmatrix}$,
(d)	$\begin{pmatrix} 0 \\ 1 \end{pmatrix}$	$\begin{pmatrix} -1 \\ 0 \end{pmatrix}$	$\begin{pmatrix} -y \\ x \end{pmatrix}$	$\begin{pmatrix} 0 & -1 \\ 1 & 0 \end{pmatrix}$,
(e)	$\begin{pmatrix} 1 \\ 0 \end{pmatrix}$	$\begin{pmatrix} 1 \\ 1 \end{pmatrix}$	$\begin{pmatrix} x+y \\ y \end{pmatrix}$	$\begin{pmatrix} 1 & 1 \\ 0 & 1 \end{pmatrix}$,
(f)	$\begin{pmatrix} 0 \\ 0 \end{pmatrix}$	$\begin{pmatrix} 0 \\ 1 \end{pmatrix}$	$\begin{pmatrix} 0 \\ y \end{pmatrix}$	$\begin{pmatrix} 0 & 0 \\ 0 & 1 \end{pmatrix}$.

2. Rectangle of dimensions 3×2; the ellipse $\frac{x_1^2}{9}+\frac{y_1^2}{4} = 25$; $\begin{pmatrix} \frac{1}{2} & 0 \\ 0 & \frac{1}{3} \end{pmatrix}$.

3. (a) Magnification with a factor of 3; (b) shear of $2x$ in the direction of Oy; (c) many–one mapping onto the line $x = 2y$; (d) reflection in the plane $z = 0$; (e) rotation through 180° about Oy; (f) reflection in the plane $y = x$.

4. Cuboid of dimensions $1\times 2\times 3$; the ellipsoid $x_1^2+\frac{y_1^2}{4}+\frac{z_1^2}{9} = 9$;

$$\begin{pmatrix} \frac{1}{2} & 0 & 0 \\ 0 & \frac{1}{3} & 0 \\ 0 & 0 & 1 \end{pmatrix}.$$

Exercise 5a

1. Det $kI = k^2$; det $kM = k^2$ det M.

2. πab; such a mapping would not be topological.

3. C_1 is the hyperbola $-25x^2+25y^2 = 1$; C is also a hyperbola.

4. (a) Singular; many–one mapping onto the line $2x+y = 0$;
(b) orthogonal; rotation through 60° about 0;
(c) orthogonal; reflection in the line $y = x \tan 30°$.

5. (a) No inverse; (b) $M_2^{-1} = \begin{pmatrix} \cos 60° & \sin 60° \\ -\sin 60° & \cos 60° \end{pmatrix}$; (c) $M_3^{-1} = M_3$.

Exercise 5b

1. $\begin{pmatrix} 0 & -1 \\ 1 & 0 \end{pmatrix}$.

2. Many–one onto the line $y+x = 0$; many–one onto the origin; X maps all points on the line $x+y = 0$ onto the origin, and all other points in the plane onto the line $x+y = 0$.

3. $M_2M_3 = \begin{pmatrix} \cos 120° & \sin 120° \\ \sin 120° & -\cos 120° \end{pmatrix}$, defining a reflection in the line $y = x \tan 60°$.

$M_3M_2 = \begin{pmatrix} 1 & 0 \\ 0 & -1 \end{pmatrix}$, defining a reflection in Ox.

$M_3^2 = I$.

6. Provided $\det M \neq 0$, $\quad M^{-1} = \dfrac{1}{\det M}\begin{pmatrix} d & -b \\ -c & a \end{pmatrix}$.

7. $D^n = \begin{pmatrix} a^n & 0 \\ 0 & b^n \end{pmatrix}$; valid for any positive or negative integer n provided $a \neq 0$, $b \neq 0$.

Exercise 6a

1. Det $kI = k^3$; det $kM = k^3$ det M.

2. $\frac{4}{3}\pi abc$; such a mapping would not be topological.

3. $x^2+2y^2+3z^2 = 1$, the equation of an ellipsoid; S is also an ellipsoid.

4. M_1 is singular, defining a many–one mapping onto the plane $2x+z = 0$; M_2 is orthogonal, defining a rotation through 30° about Oy; M_3 is singular, defining a many–one mapping onto the line $y = 0$, $x = z$.

5. $M_2^{-1} = \frac{1}{2}\begin{pmatrix} \sqrt{3} & 0 & 1 \\ 0 & 2 & 0 \\ -1 & 0 & \sqrt{3} \end{pmatrix}$.

Exercise 6b

1. $\begin{pmatrix} 0 & -1 & 0 \\ 1 & 0 & 0 \\ 0 & 0 & 1 \end{pmatrix}$.

3. X defines a mapping onto the plane $2x+2y-z = 0$; X^2 defines a mapping onto the line $-2x = y = z$; X^3 defines the mapping onto the origin. Under X, points on the plane map onto the line (which lies in the plane), and points on the line map onto the origin.

4. $D^n = \begin{pmatrix} a^n & 0 & 0 \\ 0 & b^n & 0 \\ 0 & 0 & c^n \end{pmatrix}$; valid for any positive or negative integer n provided $a \neq 0$, $b \neq 0$, $c \neq 0$.

Exercise 7a

2. $\begin{pmatrix} 1 & 0 & 0 \\ 0{\cdot}02 & 0{\cdot}10 & 0{\cdot}02 \\ 0{\cdot}01 & 0{\cdot}05 & 0{\cdot}05 \end{pmatrix}\begin{pmatrix} x \\ y \\ z \end{pmatrix}$ is a vector whose first component gives the vitamin A content of the mixture, etc.

3. $\begin{pmatrix}1&0&0&0\\0&2&0&0\\0&0&4&0\\0&0&0&8\end{pmatrix}$, $\begin{pmatrix}1&0&0&0\\0&\frac{1}{2}&0&0\\0&0&\frac{1}{4}&0\\0&0&0&\frac{1}{8}\end{pmatrix}$.

The product of the two matrices is I, as the mappings are inverses of each other.

4. $M^2 = \begin{pmatrix}0&0&2&0\\0&0&0&6\\0&0&0&0\\0&0&0&0\end{pmatrix}$. M^2 maps $p(t)$ onto its second derivative, $p''(t)$.

$M^4 = 0$.

5. $M = \begin{pmatrix}0&0&6\\\frac{1}{2}&0&0\\0&\frac{1}{3}&0\end{pmatrix}$.

Exercise 7b

2. $M = \begin{pmatrix}1&1&1&0\\0&1&0&1\\1&0&0&1\end{pmatrix}$; $M'M = \begin{pmatrix}2&1&1&1\\1&2&1&1\\1&1&1&0\\1&1&0&2\end{pmatrix}$, signifying that A lies on two routes, A and B are connected by one route, etc.; $MM' = \begin{pmatrix}3&1&1\\1&2&1\\1&1&2\end{pmatrix}$, signifying that Route 1 serves three stations, Routes 1 and 2 meet in one station, etc.

3. "Top chicken" is number 1. "Pecking order" is: 1, 2, 5, 4, 3.

4. The fourth player is the winner.

Exercise 8a

1. $E_a = \begin{pmatrix}0&0&0&1\\0&1&0&0\\0&0&1&0\\1&0&0&0\end{pmatrix}$, $E_b = \begin{pmatrix}1&0&0&0\\0&1&0&0\\0&0&5&0\\0&0&0&1\end{pmatrix}$, $E_c = \begin{pmatrix}1&0&0&0\\0&1&0&0\\0&2&1&0\\0&0&0&1\end{pmatrix}$.

2. (a) $\frac{1}{4}\begin{pmatrix}-3&2\\5&-1\end{pmatrix}$; (b) $\begin{pmatrix}-3&5&6\\-1&2&2\\1&-1&-1\end{pmatrix}$; (c) no inverse;

(d) $\frac{1}{2}\begin{pmatrix}0&3&-2&1\\2&-4&2&-4\\0&1&0&1\\-2&5&-2&3\end{pmatrix}$; (e) $\begin{pmatrix}1&0&0\\-0{\cdot}2&12{\cdot}5&-5\\0&-12{\cdot}5&25\end{pmatrix}$.

3. $T = \begin{pmatrix} 1 & 0 & 0 \\ 1 & 0 & 1 \\ 0 & 2 & 0 \end{pmatrix}$.

4. (a) $x = -9,\ y = 2,\ z = -7$; (b) $p = -12,\ q = 20,\ r = -2,\ s = -21$.

5. $5 : 34 : 25$.

Exercise 8b

1. (a) z arbitrary, $y = 1-2z$, $x = 2-z$; (b) inconsistent; (c) $a = 5$, $b = -1$, $c = 2$; (d) s arbitrary, $r = \frac{7}{4}+\frac{1}{2}s$, $q = -\frac{19}{4}-\frac{3}{2}s$, $p = -\frac{17}{4}-\frac{9}{2}s$.

2. (a) No conditions necessary; (b) $c = 2a+b$.

3. $p-2 : 6-2p : p$, for any value of p satisfying $2 \leq p \leq 3$.

6. $p = 1$: z arbitrary, $y = x = -z$; $p = 2$: z arbitrary, $y = -\frac{2}{5}z$, $x = -\frac{4}{5}z$; $p = -3$: z arbitrary, $y = -\frac{1}{5}z$, $x = \frac{3}{5}z$.

7. Any four three-dimensional vectors are linearly dependent.

8. (a) $x = -\frac{3}{2}$, $y = 9$, $z = -6$; (b) correct to one decimal place, $x = 332{\cdot}5$, $y = -1801{\cdot}3$, $z = 1724{\cdot}8$.

9. $q \neq \dfrac{-5}{4}$.

Exercise 9a

1. a, b, c are eigenvalues.

2. 4, 2 are eigenvalues.

3. Because a rectangular matrix maps a vector $\mathbf{v}$ onto an image vector $\mathbf{v}_1$ with a different number of components; thus the relation $\mathbf{v}_1 = k\mathbf{v}$ is impossible.

4. (a) Eigenvalue 3 corresponding to set of eigenvectors $\begin{pmatrix} a \\ a \end{pmatrix}$, for real a; (b) none; (c) eigenvalue 1 corresponding to set of eigenvectors $\begin{pmatrix} 6a \\ 3a \\ a \end{pmatrix}$, for real a; (d) eigenvalue 2 corresponding to set of eigenvectors $\begin{pmatrix} a \\ 0 \\ 0 \end{pmatrix}$, -2 corresponding to set $\begin{pmatrix} 0 \\ b \\ -2b \end{pmatrix}$, and 3 corresponding to set $\begin{pmatrix} 0 \\ 2c \\ c \end{pmatrix}$, for real b and c.

5. (a) $(a-d)^2+4bc < 0$; (b) $(a-d)^2+4bc = 0$.

6. $\begin{pmatrix} 2a \\ -a \\ a \end{pmatrix}$, for real a.

7. A population with an age distribution in the ratio 6 : 3 : 1 will remain the same size and in the same ratio for every subsequent year.

8. $\begin{pmatrix} 3 & 0 \\ 0 & 2 \end{pmatrix}$.

9. $\begin{pmatrix} 3 & 0 & 0 \\ 0 & 2 & 0 \\ 0 & 0 & 1 \end{pmatrix}$.

Exercise 9b

1. (a) $\begin{pmatrix} 1 & 0 \\ 0 & 0{\cdot}2 \end{pmatrix}$; (b) $\begin{pmatrix} 11 & 0 & 0 \\ 0 & -4 & 0 \\ 0 & 0 & -1 \end{pmatrix}$.

2. $M^n = \frac{1}{8}\begin{pmatrix} 1+7(0{\cdot}2)^n & 1-(0{\cdot}2)^n \\ 7-7(0{\cdot}2)^n & 7+(0{\cdot}2)^n \end{pmatrix}$. As n becomes large, $(0{\cdot}2)^n$ approaches zero, so that M^n approaches $\frac{1}{8}\begin{pmatrix} 1 & 1 \\ 7 & 7 \end{pmatrix}$.

5. $\begin{pmatrix} 9 & 0 & 0 \\ 0 & 0 & 0 \\ 0 & 0 & 0 \end{pmatrix}$.

Exercise 10a

1. Hyperbola.

2. Maximum value 27, minimum -3.

4. Ellipsoid, axes length 1, 2 and 3.

Exercise 10b

1. $\frac{2}{7}$.

3. $\frac{1}{3}$ rural, $\frac{2}{3}$ urban.

4. $\frac{2}{5}$.

5. All recessive.

6. New transition matrix is $M_1 = \begin{pmatrix} 0 & \frac{12}{5} \\ \frac{1}{3} & \frac{1}{5} \end{pmatrix}$. This has eigenvalues 1 and $-\frac{4}{5}$. By Theorem 10.2, $M_1^n\mathbf{v}$ approaches the vector $\mathbf{e} = \frac{1}{17}\begin{pmatrix} 12 \\ 5 \end{pmatrix}$ and since $M_1\mathbf{e} = \mathbf{e}$, the population will then remain static.

INDEX